Andrew McDeere

BEEKEEPING
FOR
BEGINNERS

"The definitive guide step by step to build your first hive, raise the bee colony and best to handle the honey production"

Table of contents

INTRODUCTION

Bees are insects which are strongly linked to wasps and ants, renowned for their position in pollination and, in the event of the best established bee species, western honey bee, for the production of honey and beeswax. Bees are a monophyletic group within the Apoidea superfamily and are currently regarded a clade called Anthophila. There are more than 16,000 known species of bees in seven recognized biological families. They are discovered on every continent except Antarctica, in every habitat on the planet that includes insect-pollinated flowering plants.

Some species, including honey bees, bulbs and stingless bees, live socially in colonies. Bees are adjusted for the production of nectar and pollen, the former mainly as an energy source and the latter mainly for proteins and other nutrients. Most pollen is used as larvae meat. Bee pollination is both ecologically and socially significant. The decrease in wild bees has improved the importance of pollination by commercially controlled honey bee hives.

Bees vary in quantity from small stingless bee species with a length of less than 2 millimeters (0.08 in) to Megachile pluto, the biggest leaf cutter bee population with a duration of 39 millimeters (1.54 in) for women. Halictidae or sweat bees are the most prevalent bees in the Northern Hemisphere, but they are tiny and often mistaken for wasps or insects. Vertebrate bee predators include birds such as bee-eaters; insect predators include bee-wolves and dragonflies.

Human apiculture or apiculture has been practiced for millennia, at least since ancient Egypt and ancient Greece. Besides honey and pollination, honey bees generate beeswax, royal jelly and propolis. Bees have emerged in mythology and folklore, throughout all stages of art and literature, from ancient times to the current, though mainly in the Northern Hemisphere, where apiculture is much more prevalent.

An assessment of 353 wild bee and hoverfly species across Britain from 1980 to 2013 discovered that insects were removed from a third of the sites they populated in 1980.

Honey bees, like all living things, vary in their traits across the species. Genetic differences across these breeds can lead to differences in attributes like temperament, disease resistance, productivity, color and much more. The environment has a huge impact on differences among bee colonies due to stimuli and response, but the genetic makeup of a colony is the basis for many of the characteristics that define a particular subspecies of honey bee. For as long as honey bees have been domesticated, beekeepers have known that different genetic stocks have distinctive differences that can be used to their advantage or ignored to their disadvantage. Whether it be pollination, a honey crop, bee reproduction, resiliency or otherwise, it is important to have a general grasp on what this means for you and your Beekeeping Goals.

WHAT IS BEE STOCK?A SUBSPECIES OF APIS MELLIFERA?

"Stock," as defined by David Tarpy at North Carolina State University, is a term to define a loose combination of traits that characterize a particular group of bees. Such groups can be divided by the species, race, region, population, or breeding line in a commercial operation. In many ways, the easiest way to understand Bee Stock is to compare it to the way we have followed and tracked the pedigree of racing horses throughout the ages. Often there are lines of heredity that go back hundreds of years in quality stock.

It is important to understand that although most of the honey bees for purchase come from these so-called races of honey bees, from what we know about honey bee reproduction and queen mating, the "purity" of the stock is not always easy to control. Beyond that, these races are still defined from the Old World. Dr. Al Dietz defined the terms "bee stock" and "races" as "The geographic races of bees are the results of natural selection in their homeland. That is, the bees became adjusted to their original environment, but not always to the economic requirements of beekeepers. Therefore, they are not the result, but the raw material for breeding."His statement (1992. Honey bees of the world. The Hive and the Honey Bee) helps us to understand that although these genetic strains are rooted in history, they are still just a way to help us classify basic differences and are not impervious to flaws.

THINGS TO CONSIDER WHEN DECIDING ON A BEE STOCK

There are 5 main points to consider when searching for and choosing a bee stock according to the American Bee Journal. Others take it to a a level of detail that may be beyond a new beekeeper or even a practiced beekeeper of 5 years. These types of selection criteria include:

Production

Location

Temperament

Disease resistance

Quality of products

Age of line

Reproduction Rates

Swarming Rates

Noted Range

Environmental Resilience

Much more

THE MAIN SUBSPECIES OF HONEY BEES (APIS MELLIFERA)

In the United States there are 6 main stocks of honey bees. Each strain has been studied and observed to have a variety of attributes that may be helpful to know in making your choice. Beyond that, there are local strains coming from different regions of the US and the world every year. It is always good practice to do research beyond these main strains to see if there is something that might be better suited for your area. This information comes from a variety of research institutions who have focused on making this a useful tool.

The Italian Honey Bee (Apis mellifera ligustica)

Italian Honey Bees are some of the most widely used races of honey bee stock. They originated in Italy, hence the name, and were brought to the United States in 1859. They are known for their prolific brood cycles and production, gentle nature and reluctance to swarm. As excellent producers, most commercial beekeepers will use Italians as their main source of production. They are very light colored, almost completely yellow in some colonies, making them aesthetically pleasing to the eye and fairly easy to identify.

Despite the well-rounded advantages of the Italian Bee, there are some drawbacks.

They consume resources at a rapid pace due to their long brood cycles that can last deep into the fall.

They are notorious for robbing stores in weaker or neighboring hives.

Italian Bees tend to have more difficulty with natural pests and tend to have higher collapse rates because of this. The rationale as to why this happens is yet to be determined, but research is moving quickly due to their popularity.

Italian Honey Bees are great for almost any beekeeper due to their well-rounded nature and availability in packages and nucs.

The Russian Bee (from Primorsky Krai region of Russia)

Russian bees were brought to the United States in 1997 by the USDA in response to the increase in colony collapse due to parasites.They have been noted to have natural capabilities and colony tolerance to handle varroa and tracheal mites. In fact, the US Department of Agriculture's Honey Bee Breeding, Genetics and Physiology lab in Baton Rouge, Louisiana has shown results that many stocks of this breed contain half the mite load of standard commercial stocks.

Russian bees tend to rear brood only during times of nectar and pollen flows, making this subspecies heavily reliant on the surrounding environment. Also, less availability due to constraints on breeding make this bee significantly more difficult to obtain for hobbyists or "newbees."

Beyond these traits, Russian bees exhibit some unusual behaviors in comparison to other strains. For example, Russian honey bee colonies tend to contain a queen cell almost all the time, in comparison to most stocks, where a queen cell is only present during times of swarming or queen-replacement. Another interesting trait is that although Russian colonies tend to be more aggressive, research shows that when in the presence of other strains, there is significant cross-contamination of stock and an increased susceptibility to natural pests.

Overall, this strain of bees is still being understood and is most likely not available to most beekeepers.

The Carniolan Bee (Apis mellifera carnica)

The Carniolan Bee is one of the top 2 most popular bee stocks in the United States. This strain is favored for a variety of reasons including:

Their explosive spring buildup makes it ideal for beekeepers looking to build up quickly before the summer

Carniolan Bees are extremely docile and take a lot of irritation to be agitated enough to sting.

Most notably, the Carniolan Bee has one of the longest tongues at 6.5 to 6.7 mm, which helps it pollinate crops like clover, meaning more sources of nutrition for the colony than other strains of honey bee stock.

Beyond these basic traits, due to the origin of this stock from central and Eastern Europe, these bees have been bred to be more tolerant of colder climates and rank among the best stocks for overwintering. These bees spend their winters in a tight cluster with a modest food supply and have proven to be a favorite for beekeepers in Slovenia, where beekeeping is of cultural significance.

Yet, there are some drawbacks with this strain, most notably, this bee stock tends to swarm more often than most other subspecies of bee. Some researchers attribute this to their explosive growth and comb production early in the year and even into times of dearth, as they do not require much food to survive in comparison to other bee strains.

Overall, this strain of honey bee is a great option for beekeepers who are concerned about shifting weather from warm to cold and damp or rougher winter seasons.

The Buckfast Bee (The mutt of honey bee stock)

The Buckfast Bee stock is named for the location of its hybridization and origin, Buckfast Abbey, in Devon in the United Kingdom. During the early 20th century, populations were being decimated by tracheal mites. Brother Adam (Karl Kehrle) who was in charge of beekeeping at the abbey, started to cross the strongest colonies who had survived in the area. The new stock of bees have become a favorite for those in similar environments as that of the British Isles.

The Buckfast bee shows strong resistance to some natural parasites.It has a strong knack for foraging and is not a strain that tends to swarm, making it more difficult to find these bees in the United States.Beyond this, there is often inbreeding with this strain over time. This decreases the characteristics such as resilient behavior against pests and other elements that make this aquality strain of honey bee.

The Caucasian Bee (Apis mellifera caucasica)

This bee stock was once very popular in the United States, but its lack of honey production overall has lessened its use among honey producers in the United States. Yet, there are still some commercial pollinators who use this strain due to its very long tongue; longer than Carniolan bee stock most of the time. Similar to the Carniolan bee, the Caucasian bee shares similar traits in temperament.

The lessened use of this strain in the US has increased its value among traditionalists because most stock is imported from Europe and then cross-bred with Carniolans (https://bees.wsu.edu/breeding-program/queens/). Even though the gradual build-up rate of the colony is slower than many largely used stocks, it allows for honey to be stored more efficiently near the brood. In other words, it doesn't proceed to a new comb until the previous one is completely filled.

For beekeepers who understand the importance of winter stores and the "heating blanket" in colder climates, this single trait could help your bees increase the chance of overwintering this year.

The German Bee (Apis mellifera mellifera)

The European Dark Bee, or German Dark Bee, was brought from Northern Eurasia in the colonial era. This subspecies has since then been segmented further into sub races of German Bees due to its hardiness. It is able to survive long, cold winters more often than other strains of honey bees. However, due to their defensive nature and susceptibility to brood diseases like American and European foulbrood, this stock has lost significant favor with beekeepers all over the world.

Although this strain of bee stock has lost significant value in the commercial sector, there are still researchers and hobbyists working hard to isolate the hardiness of this subspecies through tracking breeding values and data religiously.

For beekeepers in the US, this strain is most likely not available, and would not be a good choice unless you are familiar with natural diseases and parasites in the hive.

The Africanized Bee (The Misunderstood Bee)

The Africanized, or Killer Bee as most know it, is not even from Africa it originated in Brazil. This honey bee strain was a hybrid designed in a lab with the goal of increasing pest and parasite resistance, while at the same time increasing production. This bee stock showed great promise until 26 experimental swarms escaped quarantine and took over South America.

This highly aggressive strain of honey bee has some advantages, if one learns to work with them. They begin foraging at a younger age, typically produce more honey, and have a significantly smaller colony size, even though they reproduce at a faster pace. There are many stories of beekeepers working well with these bees for these positive traits.

BEGINNER BEEKEEPING STEPS WITH A TOP BAR BEE HIVE

If you are just starting beekeeping or thinking about starting, then this is the time to catch the wave to get setup for bee season. The first year of caring for bees in the top bar hive is a simple setup.

You will need to:

1) Purchase or build a hive

2) Find a resource and order bees

3) Find a location for your hive

4) Purchase protective clothing

5) Coat the top bars with beeswax

6) Install the swarm or package of bees

7) General questions for setting up a hive

1) BUY OR BUILD A BEE HIVE

Purchase a fully assembled hive from top bar Beehives for Sale page, BackYardHive, or build your own from the Bee Hive Plans on our website. They offer fully assembled hives since many folks don't have the wood shop or the time to build their own hives.

If you build a hive from their plans, I highly suggest you purchase top bars as the top bars are the most difficult part of the hive to make. They also offer a Top Bar Hive Package, which includes top bars, hive plans, spacers, and hardware. We find that having precise dimensions for the top bars encourages the bees to build straighter comb which leads to a better beekeeping experience in the hive.

They also suggest that you build the hive with the glass viewing window, this is an amazing feature that you will not want to leave out. The window allows you to observe the progress of the bees without having to disturb the colony. This also gives you the opportunity to check on your bees anytime you'd like.

2) FIND A RESOURCE FOR A SWARM, OR ORDER BEES

January or February is the time to order a package of bees or get onto a local swarm list. Bee Package Suppliers start selling their packages in the winter in preparation for a spring delivery of a package of bees. Bee Packages normally arrive in April, depending on your local climate. The weather needs to be warm enough for the bees to be put into a hive and there to be available nectar and pollen in your area for the bees to start collecting to feed their young they will start rearing. Bees start swarming in the spring. Most colonies will start swarming in April unless you are in a warmer region like the southern US and swarming can start as early as March.

3) PLACEMENT OF YOUR BEE HIVE

You will want to place the entrance of the hive away from foot traffic. The less foot traffic at the entrance of the hive the better for the bees, you and your friends. You'll need to consider the winter weather in your area and the direction of the wind. Face the hive entrance away from strong winter winds. The ideal direction would face somewhere between east and south. It is a good idea that the hive gets some shade in the afternoon in the summer and plenty of sun in the winter months. A great place for this is under a deciduous tree, where it is shaded in the afternoons in the summer, or on the east side of a building where the hive gets the warmth from the dawning sun and shade in the afternoon. You may also want to raise the hive off the ground a few feet so that it is easier to work with; a couple of cinder blocks works well for this. Level the hive, with the lid off, from front to back and from side to side. You may have to adjust the ground, or add shims under the feet of the hive.

4) PURCHASE PROTECTIVE CLOTHING FOR WORKING WITH THE HIVE.

Protective clothing means different things to different beekeepers.

When bees are being installed into the hive, they are not as aggressive or defensive as they can be once they have established. This means that they won't be concerned with stinging because they are concerned with finding their new home and colony. But you may want to wear a veil and protective clothing when first starting out with beekeeping.This helps you feel comfortable around the bees, allowing more concentration on your task.Purchase a bee suit, bee helmet and veil and protective gloves from our BackYardHive Protective Gear shop. You can also wear thick clothing, closing all arm and leg sleeves. Wrapping duck tape works well for this. And if you are in question about being allergic, you should get tested by your doctor.

5) COAT THE TOP BARS WITH MELTED DOWN BEESWAX

(paraffin will not work it needs to be natural bees wax) Coat only the 'spine' of the top bars.This encourages the bees to build comb on this 'spine'. You can order beeswax on our website or purchase it from your local art supply store.

6) INSTALL THE BEES INTO THE HIVE

Put all the top bars on the hive. Now it is time to install the bees into your hive. How many top bars you will need to remove and where you choose to put the false back, depends on the size of your swarm or colony.Typically, we suggest to insert your false back 10 bars from the front of the hive, this is important as it helps to establish the brood nest in the front of the hive.Remove 5 or more bars between the entrance and false back to install the bees into hive. If it's a package of bees the queen comes in a separate cage. Place this on top of the hive until the bees are installed in the hive. Give the box with the bees in it, a couple good stern shakes as you empty the bees into the hive.If needed give the box another stern shake, to get most of the bees into the hive.If the queen came in a cage you can put her into the hive with the bees; suspend the cage between the 3rd and 4th bars from the entrance. If you are installing a swarm make sure you locate the ball of bees on the lid of the box and carefully move the ball above the opening and give the top of the lid a stern pound, or shake. Then gently yet firmly shake the bees in the bottom part of the box into a corner, turn the box upside down and shake them into the hive. There will be bees flying all around. Hopefully, most of them are in the hive. Put all but 1 bar back on the hive.Whether it's a package of bees you bought ,or a feral swarm, you will want to make sure the queen is in the hive. They will soon start to fan, this is the bees calling out to each other "Hey, we are over here..the queen is here everyone get in!!" It can take about an hour or so for all the bees to find their way into the hive, replace your last top bar. In a couple of days, release the queen over the hive into the top of the hive , it is important that you do not drop her on the ground, so do all of your work over the hive where the top bars are removed so if she falls, she goes into her home. In 3-7 days move the false back all the way to the back of the hive, put in your spacers, and move the bees and combs 3 bars from the hive entrance, this will get the brood nest right where you want it.

WHAT HIVE TOOLS DO I NEED FOR WORKING WITH MY BEE HIVE?

Your hive tool is one of the essential pieces of beekeeping equipment you will need to work with your hive.If you only get one tool the ultimate Hive Toolspecially designed by Corwin Bell The ultimate Hive Tool that is versatile and the best choice for working your bee hive whether it is a top bar hive, Warre or Langstroth. An all-in-one sturdy hive tool for easily removing combs and prying apart top bars when working with your hive.

How Many Hives Can I Start At Once?

You can start as man hives as you feel comfortable managing. If you can build or buy more than hive it is a great idea to start with more than one hive as your learning curve doubles with 2 or more hives. Each hive will have a different personality and will have different traits like how fast they build combs, how much they guard the hive, how much they use propolis in their hive, the list goes on

How Far Apart Should My Bee Hives Be?

It is perfectly fine to put the hives in close proximity. Here are some things to think about.

1. You want enough room to walk between the hives and be able to work around each hive. The biggest problem for me has been when I am harvesting honey in one hive and the adjacent hives' bees catch wind of the honey smell. I now have increased the number of bees landing on the combs I am trying to work. This is one reason I like to do the the single comb harvest with my hives that are close together. I can go in and remove a comb before any of the "other" bees get what's going on.

2. If the hives are close together they get used to each others' proximity and scent and let their guard down, which sometimes results in them robbing the other hive. Disease can more easily spread and infect hives in close proximity; this is much of the problem with big honey apiaries. I like to give the bees their own local territory. If you have the room I would put the hives several hundred feet away from each other. Otherwise 4 to 5 feet or enough room to walk between them and still be able to work the hive.

3. I get different honey from these hives.

A scenario might be; my #1 hive is right next to an apple tree will be a dark "apple honey" and the #2 hive, in the side-yard is going after the melon patch and the plum bushes across the road.

One of my favorite bee keeping joys is seeking out mono-flora areas. I drive and look for sunflower fields, different flowers down by a lake or fields with wild flowers. I will then put a hive in this area for a couple of months just to see what the flavor and color of the honey is.

Can I Install Two Hives At The Same Time?

If the hives are thirty feet away or more, you could install them at the same time, although thirty feet is really an arbitrary number - the main thing to consider is the time when the bees formed their colony and regrouped. When the first bees installed have stopped "fanning", then you can install the second colony of bees. Fanning is pretty neat to watch. The bees that are around the new hive will put their abdomens up in the air and rapidly fan the air. The bees are putting out pheromones and sonic vibrations signaling their buddies where the queen and the new hive are. So you can imagine if two colonies are doing this at the same time, it could be a little confusing.

1. Choose the location Bees need four things. First, they need sun, or afternoon shade if your weather is hot. Second, they must have access to fresh water near the hive. We used a large plant saucer with stones in the center for the bees to land on and refreshed the water every day. A shallow bubble fountain would work well, too. Third, the hive must be protected from wind, which can blow rain (or snow) into the hive, making it harder for the bees to keep the hive warm. Finally, bees need privacy. Don't put the hives near high-traffic areas, play areas, swimming pools, or pet areas. Give each hive plenty of space 50 feet from high-traffic areas is ideal, but if space is limited, position the hive so the entrance is near a tall fence or hedge. This will force their flight path overhead to minimize contact with people and pets. And screening them from view will keep bees and people happy.

2. Prepare the location Hives should face south, if possible, and they need to be kept off the ground to protect them from dampness and critters. After clearing the brush and leveling the ground, we poured a cement pad to make care easier.

3. Install the bees Spring, when blooming flowers furnish a food supply, is the time to put your bees in their hives. Once you've chosen how to buy them, the best bet is to rely on your source for installation instructions.

Here is what happened when we picked up our bees from Randy Oliver at his property in Grass Valley: He gave us an introductory class in beekeeping, showing us how to use the hive tool and the smoker, handle bees, and check for eggs, brood (larvae), and queen all vital signs of a colony's health. Randy loaded 5 frames of his gentle hybrid bees and a queen into each of our two brood boxes and sealed the openings by stuffing them with our beekeeping gloves. We used ratchet straps to secure the boxes in the back of our truck. When we got back to Sunset, we positioned the brood boxes in their designated locations and removed the gloves from the entrances. One hive we named Betty; the other, Veronica.

4. Feed the bees Young colonies have a lot of work to do—storing pollen and nectar, sealing all the cracks and seams in their new home, and taking care of the queen and new brood. To make their adjustment easier, we fed them a "nectar." Here is how to make it: Dissolve equal parts granulated sugar and water and use to fill the quart jars. Top with the feeder lids and invert the jars into the holes. The lids should not drip; they should be barely moist. The bees will drink what they need from the lids.

In the beginning, our nucs drank about three-fourths of a quart jar per day. Over the next 3 weeks or so, it tapered off to the point where we realized sugar water was no longer necessary. The bees were finding their nutrition in flowers. Plus, sugar water makes for insipid honey and should not be continued if it is not needed.

5. Inspect the hives inside and out Much of beekeeping is simple observation and response. If you are a novice beekeeper, inspect the hive about once a week for a couple of months so that you can learn. Once you feel comfortable, adjust your routine to every two weeks. Make sure the outside of the hive is clean and free of bee poop, the landing board is free of litter, and there are no ants on the hive. Open the hives and check frames for larvae and eggs (on warm days only). If the queen is healthy, you will see plenty of larvae in various stages of development.

If you don't see evidence of a healthy queen, consult an expert. Your local beekeeping guild is a good source.

Ultimately, the less often you inspect the hive, the better for its health. Opening the hives and thoroughly checking them requires smoking to keep the bees calm. This stresses the bees and it takes them about a day to recover. As you learn more, you will find you won't need to pull many frames to know what is going on inside. And you will figure out a lot simply by observing the bees as they come and go from the hive.

6. Check regularly for pests and diseases Varroa mites are the pest most typically found in hives. Left unchecked, they can cripple and eventually kill the hive (see Pest Control, below, for hints about checking for mites and mite control). Other pests you need to watch for include the small hive beetle and the wax moth. Diseases you need to be on the lookout for are American and European

foulbrood. Early intervention can often mean the difference between a healthy hive and a dead hive.

7. Expand the hive when necessary Start with one deep hive body-brood box. When the bees have filled it with 7 or 8 frames of bees and brood, top it with a second brood box. Let the bees build up brood cells in the second brood box, too. When the second brood box is well filled (7 or 8 frames of bees), top it with a queen excluder, if you choose to use one, and, finally, the honey super (the box from which you will collect most of your honey).

Flying honeybees

A hive with a low colony count may struggle to defend invasions that can occur via large openings. These weak hives can benefit from entrance reducers.Beekeepers use t hese small blocks of wood to protect the hive from robbing honey bees or yellow jackets. While these add-ons aren't reccomended during the height of a honey flow, they can be helpful during winter to discourage drafts, rain, snow, and mice from entering the hive.

PEST CONTROL

Bees are like flying balls of delicate spun sugar filled with honey. Everything wants to eat them. Here are three of the worst pests we battled, and the tactics we used.

Ants Argentine ants can kill a hive by robbing honey and eating the brood. We couldn't spray to kill the ants, since that would also kill the bees. We tried Terro ant bait little containers filled with boric acid mixed with a sweet substance ants like with some success. In the end, we were most successful with a physical barrier. We placed each leg of the hive stands in plastic tubs filled with water that the ants could not cross.

Small hive beetles Hive beetle larvae will eat all parts of the hive, including the baby bees. We kill the beetles on site, and have been experimenting with traps like AJs Beetle Eater ($5.25) from Dadant.

Varroa mites The most damaging pests a beekeeper has to deal with are these mites, as they threaten the survival of a hive once they become established. They suck the blood of adult bees and lay their eggs in brood cells, where their larvae feed off bee babies, infecting them with viruses and weakening and even killing them. To save their bees, beekeepers use a variety of methods:

1. Monitoring A 24-hour count of a natural mite fall will give you a good idea of a hive's infestation. Coat the bottom of your Country Rube board with petroleum jelly or cooking spray (to trap the mites), slide it into the lower part of the bottom board, wait for 24

hours, and then pull it out and count the mites. Anything more than 10 mites per brood box indicates you have a problem.

2. Sugar dusting The powdered sugar method lets you both count the mites and control them. Sift powdered sugar, 1 cup per brood box, over the tops of the frames and brush it into the hive. The powdered sugar makes the mites lose their grip on the bees and fall off; plus the bees groom the sugar off their bodies, dislodging more mites. Again, use the bottom board to capture the fallen mites. You should not see more than a few mites 10 minutes after dusting. If there are more, you have a problem.

3. Mite trapping Drone frames will also help trap varroa mites. These frames are designed to encourage bees to make drone comb cells, which are larger than worker comb cells. Since varroa mites prefer drone brood 10 to 1, the drone comb makes a great mite trap. Just before the drones hatch (24 days after the eggs were laid), destroy the drone comb (you can freeze it and return it to the hive, or simply cut it out), and replace the drone frame for the next cycle. (Since our queens have already mated and have a lifetime's supply of sperm inside of them, they do not need the drones in order to reproduce.)

4. Apiguard A gel infused with thymol, made from the oils of thyme plants. It works well, but it makes the honey stored during the treatment taste like mouthwash.

5. Formic acid More toxic than thymol, formic acid kills the mites by gassing them. It makes the honey inedible for humans, so it is applied in the fall and winter, when the nectar flow is slow or stopped. You need to wear a respirator when applying it.

HONEY COLLECTION

We were lucky to collect honey the first summer. Typically, during the first year the bees build up their hive, and if they overwinter well, you can begin harvesting in the late spring or early summer of the second year. Three months after bringing our bees home, we had 4 frames packed with honey, each weighing about 8 pounds. Lacking a professional extractor, we used the following low-tech method.

1. Cut and crush Using the bench scraper, we cut the honey wax and all off the foundation into a bowl, balancing the frame on a wooden spoon set across the bowl like a bridge. Then we used a wooden spoon to crush the honey and wax in the bowl.

2. Straining and settling We poured this slurry of wax and honey through a double layer of cheesecloth and the stainless-steel strainer into our food-grade plastic bucket. Then we left it to drain and settle for a couple of days (bubbles and foam rose to the surface).

3. Bottling We covered the floor with newspapers and got our jars ready. Then we loosened the honey gate (the stopper at the bottom of the bucket) to release the honey into each jar.In went the honey, on went the lids.It was as simple as that. From 4 full frames of honeycomb, we reaped 12 pounds, 10 ounces of honey. We rinsed the leftover wax and froze it. Later, we rendered the wax in a solar wax melter and used it for craft projects like lip balm and hand

salve. We had a second surprise harvest later in the summer, bringing our total to about 31 pounds of pure, fragrant honey.

HONEY BEES FOR SALE

A FEW POINTS TO CONSIDER BEFORE YOU BUY

If you are starting beekeeping, no doubt you will be looking out for honey bees for sale. But how should you make your choice?

Firstly, if you are establishing a colony, you will probably purchase a nucleus colony or a 'nuc'. A nuc consists of a queen, brood, and workers in four or more frames.

Alternatively you may purchase package bees, which will consist of workers and a newly mated queen, and food.

Before purchasing, here are a few considerations:

- The bees should of course be disease free.

- Where do they come from – which race/sub-species of honey bee are they? Some suppliers import queens from overseas, but some beekeepers prefer to purchase queens reared in their own country to minimise the risk of bringing in new strains of virus or disease. Ask the supplier for details.

- Temperament - if this is your first time beekeeping, and you are nervous, then you may prefer to choose 'mild mannered' bees.

Note there are additional circumstances when you may wish to purchase honey bees, such as the replacement of winter losses.

More considerations

Timing

Note that if you wish to buy bees, you may have to order in advance so that you don't miss out. Ensure you have everything set up to and ready for your bees before you receive them. Also, have a good supply of food for the bees, until they settle and begin their normal foraging.

Take a look at my lists of bee plants to ensure you have a good range of nectar and pollen available in your garden for the bees throughout the seasons. If you live in an urban area, there may be plenty of flowering plants in the vicinity, from gardens and public planting schemes.

Swarms

What if you are offered a swarm? If you are a newcomer to beekeeping, you may want to get help from another beekeeper to settle in the swarm into your hive.

Where Else Can I Get Honey Bees?

- Make contacts with your local beekeeping association, who may be alerted to the locations of swarms that need to be moved.

- Contact your local council, and let them know you are wanting a swarm. If a member of the public asks to have a swarm removed, they can then contact you.

WHERE TO GET BEES

Packaged bees and caged queen It takes time to build up the colony this way, but it's the least-expensive choice. You can usually order packaged bees through your local beekeepers' guild.Preorder as early as the fall and certainly no later than early spring, as bees are only available for a short time in spring.About $65.

Nuc (short for "nucleus") A nuc is a young hive, usually covering no more than 5 frames of comb, with a newly laying queen. Starting this way helps you get a jump on honey production.Buy from a reputable beekeeper to avoid getting diseased equipment or sick bees.We ordered two nucs from master beekeeper Randy Oliver and drove to his location in Grass Valley, California, to pick them up.$90 for each nuc, queen included; scientificbeekeeping.com.

Well established swarms or colonies Large colonies can be daunting if you've never kept bees before, and beginning beekeepers shouldn't try to capture a swarm. Leave that to a more experienced beekeeper (contact your local beekeepers' guild to find such a person), and perhaps he or she will help you start a hive with the captured swarm.

BEE ACTIVITY BY SEASON

Each season brings its own challenges to a hive and the bees react accordingly. The activity level of a bee is dictated by the weather and cold weather inhibits their movement. In fact, bees left out in the elements, unprotected will often not survive the winter.

A high-level concentration of bees in a beehive will be better able to tolerate the cold than a hive with fewer bees. But any hive is in danger of succumbing to the cold if the temperature drops too low. Your role is to act as custodian of the hive, keeping it safe and in good condition throughout the year. The rest is up to the bees!

Winter

During the winter, the bees will form a tight cluster around the queen to ensure her survival and keep each other as protected as possible during the coldest months. Only female worker bees will be in the hive at this point, having forced out the drones who no longer serve any purpose.

The bees will eat about 50 pounds of honey during this period, but will run out of food before the winter weather eases. You will need to provide the hive with sugar to supplement their appetites toward the end of the season. In the meantime, your main objective is to keep snow and ice off the hive and ensure there is an emergency food supply available.

Of course, since bees will be confined to a hive, starting a beehive in the winter makes no sense.

Spring

Honeybees will often begin spring close to starvation, and so you will often need to provide an emergency food supply until the flowers begin to bloom and the bees begin to harvest their own food supply again.

If you are starting a beehive in the spring, it is generally necessary to help the bees by feeding them sugar syrup. This will help them build resources inside the hive until they are able to find a supply of foraging resources in the area, at which point they will hopefully become self-sufficient.

As the weather warms up, the queen will more produce eggs and their food stores will gradually replenish. You should monitor the bees to make sure that everything is going according to plan and that they are producing adequate food for themselves. It may be necessary to supply them with them with emergency food rations until you observe that they have an adequate food source of their own.

By May, the activity in the hive should be in full swing and the drones that were eliminated when winter began will have been largely replaced. You should maintain your normal beekeeping routines at this time, including medicating the colony and adding more supers as the need for them arises. This period may see an explosion in the hive population, so be sure to keep up with the maintenance of the hive.

Summer

June and July see the worker bees in constant motion collecting pollen, producing honey and tending to the queen. The drones will be thickest through this period, but you will see their population begin to diminish in August as their usefulness to the queen wanes.

Be mindful of predators and other insects that may be attracted to the honey and may cause damage to the hive or the bees. As September begins, the drones' presence will drastically change and you will see the population of the hive decline as the drones are dismissed by the worker bees.

Fall

At this point, you will probably have harvested the hive's honey, but be sure to leave around 60-70 pounds of honey for the bees to survive the harsh months ahead. The queen will lay fewer eggs now as the workers' activity begins to die down.

This is when the beekeeper should medicate the bees and supplement their food with sugar syrup. As fall progresses, the bees will begin to huddle around the queen and there will be little activity from them. You should use this time to reinforce the security of the hive to keep out rodents and predators that will make a meal out of a hibernating brood colony.

November and December will call for little more than cursory inspections to make sure that everything is as it should be.

Weather Conditions for a New Hive

When spring begins and the weather begins to warm is the ideal time for bees to start a new hive. Their activity levels build along with the blooming of flowers and they can tend to and build their hive as the season flourishes. Cold weather will inhibit the bees' movements and they will not work on the hive until the season changes. Spring may begin later in some regions than others, so you should plan for your specific region.

Timing to Start in the Spring

When starting a beehive, the spring is ideal time. As soon as the weather begins to warm up and flowers start to bloom, you can set up your hive. Plan well and get your bees into the hive as soon as possible to allow them the time to gather as much nectar as possible through the next few months. By the time fall rolls around, the hive should be well established and honey should have been produced in abundance, barring any unforeseen complications.

When To Order Bees

You need to order your bees to arrive in early spring to get them installed in the hive on time. Ordering and receipt of your bees are two different things! Be sure to order well in advance to ensure your bees do indeed arrive on time. Talk to local beekeepers to identify the optimum time to order from local suppliers.

Don't jump the gun and accept delivery before the weather warms up in your area (obviously this differs across the country). You want the bees roused enough to begin building the hive and making it their own quickly. For beginners, do careful research on when your region will ring in the spring fully and prepare for your hive from there.

FEEDING HONEY BEES FOR BEGINNERS

In my first year of beekeeping. I was surprised to learn how often I had to feed my new pets. I wasn't supposed to be amazed, like all the livestock that honey bees have to consume.

Unfortunately, I've done a lot of reading on the Internet and heard that feeding your bees is bad–unnatural, unhealthful, makes them lazy, and swarm, can make them produce brood at the wrong time, etc, etc... Anyway, if you don't take too much honey out of them, they won't have to be fed. Well, that last declaration may be true–in a nice year, with a well developed nest. But not in most years in Mid TN, and definitely not for the first year of hives that have not yet been developed.

In My Opinion, ensuring that your bees–or your dog, bird, or children–have a good nutritional year around is essential to maintaining them healthy and efficient.

Yes natural fruit and honey may be more perfect plant food than milk and pollen, but inform me that you'd let your children starve because new fruit is out of season and canned beans are not perfect? Well, how about your dog? There is not a lot of dog chow accessible in nature?

So assuming that we agree that you sometimes need to feed your bees, we're mostly speaking about simple ancient table sugar in one shape or another. Store on it, and watch for sales 4-5 pound packages are often cheaper, and easier than bulk sugar until you have a few hives. But always hold a few of them on side just in emergencies.

How do you understand when you want to feed?

Very simple. Do your inspections. Any hive that does not produce at least 15 pounds of honey and free flowers mixed with nice quantities of both should be supplied. Starting around mid-September, this amount shifts to 50 pounds of healed "honey"and strong sugar. Once those things are achieved, you can avoid feeding until they use some of them and need to be supplied again. Which you'll understand because you're supposed to do the inspection.

While doing your checks, you need to look out for the backfilling of the brood nest if you see the flowers placed in the brood nest, then stop eating and stay vigilant. Allowing backfilling to proceed either because you're eating, or because of a natural water stream will lead in swarming if it lasts for just a few days.   In addition to not eating, you can also offer them more blank drawn comb to function with or open the family shed by inserting a baseless frame the comb will be built in it and the woman will put eggs in the fresh comb. Both of these operations assist stop swarming.

Luckily, there is usually plenty of pollen available in our area almost any time the bees can forage so you don't really need to worry about using pollen replacements in your first year. But 1 caution just in case you decide to test the hot climate pollen sub particularly in the shape of "patties" is generally very rapidly infested with insect larvae.

What kind of feeder should I use?

Inverted quart containers or in the drop when I use reversed 2 gallon pails so I can get the work accomplished a lot faster. Cheap. Don't destroy the bees. They're also putting the feed directly on top of the bees where they can get it even in hot climates. You can either put it on top of the hive if it has a hole in the cover how I do it over a hole in the inner cover, enclosed in an empty super, or directly over the top bars. If you're using jars, don't let the sun shine on them heat the product and spread it to the hive cover it with an open super, coffee can, aluminum foil anything to hold the immediate sun off.

Miller feeders are popular easy to complete, and they retain a ton of food. Expensive and generally some bees drown. If you put them on during the summer, they may be stuffed with a comb.

Boardman feeders–not advised because they can break the robbery off.

Frame feeders they need to open the nest to eat, swallow some birds but they're putting the feed where they can get it even in hot climates.

Ziplock baggies–cheap one-use solution–don't destroy bees if they're finished properly. They don't have a lot of food, and they need to open the hive to change you can't refill it.

WHAT TO FEED BEEN IN THE HIVE

Honey bees store honey in the hive to provide food for winter and for other times when there are few or no nectar secreting flowers available.

When nectar is in short supply or unavailable, bees draw on their honey stores in the hive. During these times, it is important to frequently monitor the amount of honey in the hive because when it has all gone, the colony will starve.

Starvation can be prevented by moving bees to an area where plants are yielding nectar or by feeding them white table sugar, or syrup made with white sugar. Never use raw, brown and dark brown sugar, and molasses as these may cause dysentery in bees. Bee colonies can be kept alive for long periods by feeding white sugar.

HONEY AS FEED FOR BEES

It is extremely important not to feed honey to bees unless it is from your own disease free hives. Spores of American foulbrood disease can be present in honey. Feeding honey from an unknown source, for example, a supermarket or even another beekeeper, may cause infection in your hives. If you feed suitable honey to your bees, it must be placed inside the hive. Never place honey in the open outside the hive as this is illegal under the Livestock Disease Control Act 1994.

How And When To Feed Bees

If sugar syrup or dry sugar is fed in the open, bees from nearby managed and feral colonies will be attracted. You will end up feeding other bees as well as your own. Besides being a waste of money, feeding in the open may cause robber bee activity in the apiary and possible interchange of bee disease pathogens.

Placement of sugar syrup or dry sugar in hives is best done towards evening to minimise any tendency for bees to rob the hives that are fed.

MAKING AND FEEDING SUGAR SYRUP

There are differing views about the correct amount of sugar in syrup. Some beekeepers prefer a ratio of one part of sugar to one part of water, measured by weight (known as 1:1). Others prefer a dense syrup of two parts of sugar to one part of water (known as 2:1). Generally, 1:1 syrup is used to supplement honey stores, stimulate colonies to rear brood and encourage drawing of comb foundation particularly in spring. The stronger syrup is used to provide food when honey stores in the hive are low. Measuring the sugar and water by weight or volume is alright because there is no need to be 100% exact about the sugar concentration.

Heat the water in a container large enough to hold both the water and sugar. As soon as the water has begun to boil gently, remove the container from the heat source. Pour in the sugar and stir the mixture until the sugar crystals are dissolved. Never boil the mixture as the sugars may caramelise and may be partially indigestible and toxic to bees.

Always let the syrup cool to room temperature before feeding it to bees. The cooled syrup can be given to hives using one of the following four methods.

Container with sealable lid

Fill a clean jar, tin with a push-down lid, or similar container with sugar syrup. Drill or punch the lid with 6-8 very small holes. Cut two 12 mm high risers from a piece of wood and place them across the top bars of the frames that are in the top box of the hive. Invert the filled container and place it on the risers. Next, place an empty super on the hive to enclose the feeder and then replace the hive lid. The risers provide a bee space between the top bars and the holes in the container lid. It is a good idea to remove the cardboard insert commonly found in jar lids.

Plastic bag

Partially fill a plastic freezer bag with sugar syrup, about half full. Gently squeeze the bag to expel all the air. Tie the neck of the bag using an elastic band. Place the bag on the top bars of the frames in the top box of the hive, under the hive cover. Use a brad or very small diameter nail to punch about 6-8 small holes into the upper surface of the bag. The bees will suck the syrup through the holes. Never put the holes on the under surface of the bag as the syrup may leak out faster than the bees can gather it. This may lead to loss of syrup outside the hive and cause robbing by nearby bees. It is important to have a bee space between the upper surface of the bag and the under surface of the hive lid so the bees can gain access to the syrup. If required, a wooden riser of the dimensions of the hive may be used to raise the lid.

Shallow tray

Place sugar syrup in a shallow tray, such as aluminium foil tray, under the hive lid. Bees need to be able to reach the syrup without falling into the liquid and drowning. Some grass straw or wood straw of the type used in cooling devices may be placed in the syrup for this purpose. It is important not use any straw or floating that has been treated, or been in contact, with chemicals as this may be hazardous to bees. The hives should be on level ground to prevent loss of syrup and a riser may need to be used if the tray is not shallow.

Frame feeder

Place sugar syrup in a 'frame or division board feeder'. This is a container, the size of a full-depth Langstroth frame, that has an open top and which sits in the super as a normal frame does. The feeder requires a flotation material or other means to allow bees to access the syrup without drowning.

How Often To Feed

It is normal for bees to remove syrup from a feeder, reduce the water content and store it in the combs as if it were honey. Whatever feeder is used, a medium to strong colony will usually empty it in a matter of days.

For colonies with virtually no stored honey and no incoming nectar, the initial feed will be largely determined by the amount of brood, the size of the colony and to some degree, the size of the container used to hold the syrup. It is safer to over-feed a colony than to skimp and possibly cause the death of the colony. Some beginners have tried tablespoons of syrup, but this amount is much too small. An initial feed of around 1-3 litres could be tried. It is then important to frequently check the combs to see how much syrup has been stored. This will give a guide as to how often and how much syrup should be given. Feeding can be stopped when nectar becomes available.

Properly ripened syrup should have a moisture content of around 18%. Syrup that is not ripened adequately will ferment and adversely affect bees. Colonies with insufficient stores for winter should be given enough syrup to boost their stores before the cold weather of autumn sets in. This will enable the bees to fully process the syrup.

Feeding dry sugar

Medium to strong colonies can also be fed dry white table sugar placed on hive mats or in trays under the hive lid. Bees require water for liquefying the sugar crystals. They will obtain supplies from sources outside the hive and sometimes use condensation that may occur inside the hive. Some beekeepers prefer to wet the sugar with water to prevent it from solidifying. In effect, this creates a partial syrup. Weak colonies may be incapable of gathering sufficient water and feeding of dry sugar to them is not recommended. Regardless of colony size, feeding dry sugar works best during autumn and spring when humidity is relatively high. The hot, dry conditions of summer make it hard for bees to dissolve sugar crystals into a liquid.

It may be preferable for a colony at starvation level to be first fed syrup before dry sugar is given. This will give the bees immediate food without the need to liquefy crystals. Bees will generally not use dry sugar when they are able to collect sufficient nectar for the colony's needs. The sugar will remain in the hive and in some cases will be deposited by the bees outside the hive entrance. A small amount of dry sugar may be converted to liquid and stored in the cells.

Important note

Sugar remaining in combs must not be extracted with the next honey crop. The sugar will contaminate the honey and the extracted product will not conform to the legal standards set out in the Australia New Zealand Food Standards Code - Standard 2.8.2 – Honey. Ideally, the amount of sugar that was given to the hive will be fully eaten by the bees at the time hives are placed on a honey flow. This is not always possible to achieve. Also, during expansion of the brood nest, sugar stored in brood nest combsmay be moved by the bees to the honey super.

When To Stop Feeding Bees

Never feed bees sugar water when honey collection supers are on the hive. I am referring to boxes of honey that are intended for human consumption.

The bees will use any nectar (or nectar-like substance) to make honey. Honey or honeycomb produced from sugar water instead of nectar that's a no no. And, its not real honey.

Established colonies can usually survive on their own unless you are in a drought. New colonies need extra help.

BEEKEEPING KITS WITH JAR FEEDERS

Each year, new beekeepers begin their adventure with beekeeping kits. Perhaps, they ordered the kit for themselves or maybe it was a gift.

Beekeeping kits can be a good purchase depending on the quality of items included. Beware of beekeeping kits that come with small jar feeders.

The small jar bee feeders can be a hazard in the hands of inexperienced beekeepers. They make look cute hanging on the front of the hive but should only be used inside the hive.

Let me repeat that as it is very important Jar Beehive Feeders should only be used INSIDE a hive. This is especially true if you have more than 1 hive in the location.

A kit without a feeder is a better value for most new beekeepers. Then, you can choose the style of feeder that you would like to use.

Jar Bee Feeders you Need More than 1

Quart jar feeders can be done with materials at home or you can purchase the jar holders also called Entrance Feeders or Boardman Feeders.

If made at home, sugar water is mixed well and poured into the glass jars. Granulated cane sugar is the sugar of choice when making bee sugar water.

Using a small nail, punch several tiny holes in the metal jar lid. When inverted and placed on the hive, bees will access sugar water through the small holes.

If placed directly on the top bars of the hive, you will need an extra deep box to enclose the feeder.

Some beekeepers use a temporary hive top with a 1″ – 2″ hole in the top. (You can use a wide drill bit to do this.) The upside-down jar is placed over the hole.

Be sure to secure the jar to the hive in some way.i.e. Cover it with a plastic bucket and place a brick on top etc.

We do not want the wind or a raccoon to push the jar over. Of course, you can also use an empty deep box (with a top) to enclose jars inside the hive.

A Tip for Jar Bee Feeders

One easy way to use jar feeders is to purchase or make 4 jars feeders. Place all 4 inside the hive sitting on the inner cover.

Use an extra deep hive box to enclose the jars and then put the top back on the hive. This allows the bees to have access to a gallon of sugar water inside the hive.

Bees Should Not Need Constant Feeding

Nectar availability ebbs and flows in most areas. It is subject to climate conditions such as drought or a late freeze.

Strong winds, cold temperatures or rain can affect foraging conditions. If the population of your beehive is small, you do not have as many workers to collect nectar.

Too little food during Spring build up causes the bees to sacrifice brood or baby bees. Poor foraging conditions in the Fall prevents storage of food for Winter survival.

The bottom line is that we cant always rely on natural nectar. This is especially true because we keep more hives in a one area than you would find in nature.

Feeding Honey Bees Sugar Water for New Colonies

Let's talk about package bees. The majority of new bee hives established each year begin with package bees.

This small wooden and screen box contains about 10,000 bees and a queen. Other than a small can of sugar water, the bees have no resources but themselves.

Experienced beekeepers often freeze excess frames of honey. These are fed back to colonies at a later time. Unfortunately, the new beekeeper does not have this option.

Feeding honey bees sugar water is necessary in most areas. Checkout my post on feeding package bees .

What to Use for Bee Syrup or Sugar Water

I prefer sugar water made with pure cane sugar. Avoid powered sugar, brown sugar, molasses etc when feeding bees. These sugars have indigestible components that can make your bees sick.

If the bag does not say pure cane sugar, it probably is not. You should be able to find pure cane sugar in your local grocery but you can order it if you prefer.

Some beekeepers, especially those with a large number of hives use regular granulated sugar. It is most likely made from beets. This should work well as far as we know there are some concerns about beet sugar. How to Make Sugar

#1 Problem New Beekeepers Face When Feeding Honey Bees Sugar WaterThere are 2 basic recipes for sugar water used by beekeepers.

 We call them 1:1 (1 to 1) and 2:1 (2 to 1) ratio.

Let's break this down because to a new beekeeper it can be so confusing.

You can measure by weight or volume it does not matter. New colonies are fed 1:1 to encourage brood rearing.   2:1 (2 parts sugar to 1 part water) is used in the fall to encourage storage of food for winter.

Sugar water honey will keep your colony from starvation but it is not real honey. A golden rule to remember when feeding bees sugar water.

Remember, never feed bees when honey supers on your colony are meant for human consumption.

Many beekeepers use a feeding supplement in their sugar water. It promotes good feeding and prevents your syrup from becoming moldy.

Containing essential oils, these products are reported to promote good health in our bee colonies.

These products are concentrated add only a small amount to the sugar water for your bees.

Honey Bee Feeders

A multitude of feeders are available for feeding honey bees sugar water. Each type or method has pros and cons. Lets explore the most common bee feeders.

A Boardman Feeder (Entrance Feeder)

Use this feeder with a regular glass canning jar. It fits into the front hive entrance. A beekeeper can easily see when to refill.

But, this type of feeder does have its problems. It does not hold a lot of sugar water and will require refilling often. A hungry colony can drain this in a couple of hours.

The smell will attract wasps and bees from other hives to the entrance. This may result in fighting and robbing.

If you want to use a boardman feeder, place it inside the hive. The entrance feeder can be placed on top of the frames needs an empty box around it to close up the hive. You can use several feeders at one time!

The Hive Top Feeder

Hive top feeders sit on top of the hive under the telescoping top. They may be made of wood or plastic. This is a great option for the new beekeeper who is not able to visit their hives daily.

The feeder will hold around 1-2 gallons of bee syrup and will feed the colony for several days. One disadvantage of hive top feeders is their tendency to leak over time.

They are also heavy when filled and care must be taken to avoid spilling syrup around the hive.

Frame Feeders

A frame feeder takes the place of one frame in the hive body. Most frame feeders hold approximately 1 gallon of syrup. Being inside the hive, the bees have easy access to the food.

Frame feeders have some cons. You have to open the hive to refill. And, you will always have some bees drown in the feeder.

If you choose to use frame feeders, put some type of floating material inside. I have used small sticks or wooden Popsicle sticks to reduce bee drowning.

Open Feeding Sugar Water

This methods for feeding bees has some merits but it is also risky. Honey bees drown very easy. Also, it is not an economical way to feed. You are feeding every bee and wasp in the neighborhood.

If you do plan to try open feeding, ensure that the feeders are well away from your hives. At least 50 feet, more is better or you increase the risk of creating a robbing frenzy.

Bucket Feeders (Or Pail Feeders)

The bucket feeder is one of the most popular bee feeders. A small plastic bucket with a mesh feeder hole holds 1 gallon of sugar syrup. (You will also find larger styles available.)

Fill the bucket with sugar syrup. Put the lid on tightly. When you turn the bucket upside down, some syrup will escape until a vacuum forms.

Put the upside down bucket directly on top of the frames. Bees will feed from the mesh feeder hole in the bucket (or small holes drilled by the manufacturer).

Like the boardman feeders, this method requires extra equipment. An empty hive body around the bucket allows the hive to be closed.

If you choose, you can use a temporary hive top with a small hole drilled in the center.

Place your upturned bucket over the hole. A brick or rock on top of the bucket will prevent wind damage.

Refilling the bucket can be accomplished easily without disturbing the colony. And no drowned bees !

Outside feeding is interesting to watch but it causes frenzy activity. Never use outside feeders near your hives!

The biggest mistake made by new beekeepers is failing to feed a new colony long enough. Feeding honey bees sugar water can not have a set time frame of when to stop.

When bees first arrive, they are hungry and feed heavily. In a few weeks, a natural nectar flow (or honey flow) may occur in your area.

When attempting to buy honey bees, we often want bees that will be delivered right before this time.

A beekeeper who stops feeding in April may find a surprise in the September hive. No food stores for winter and frames that do not have honeycomb drawn out are common.

The new colony full of potential in the spring now faces starvation.

The moral of this story is feed your bees. If they lose interest in the sugar syrup, remove it for a few weeks. Then offer a small amount of food. If they take it quickly, you will know it is time to start feeding again.

You should stop once they colony has drawn out the comb in all your frames and filled some with honey. (How much will depend on your climate. In my area, I want a deep and a shallow full by September.

FEEDING SUGAR WATER IN WINTER – NO

If you have colonies that need additional Winter stores, feed them while the weather is warm. Bees can not make good use of sugar syrup in cold temperatures.

Preventing Robbing When Feeding Bees

I love my bees but they are all little "robbers and thieves". Especially during times of a nectar shortage or complete lack of nectar (we call that a dearth)you may experience robbing.

The stronger or more aggressive hives with fight the weaker hive and steal honey(sugar water). Once this begins it is difficult to stop.

If you see a fighting frenzy at the front of your hive, cover the hive with a damp towel. The towel hanging down loosely over the entrance of the hive tends to calm the situation.

If that is not successful, I have turned a water sprinkler on the hive for a few hours.

Robbing is best avoided. Be vigilant when feeding bees, keep entrances small enough for the colony to defend.

If a small colony, entrance should be only about 1 inch, a larger colony of strong nuc can have a bit larger opening.

Feeding bees is a lot of work and expense. How much you need to feed will depend on your climate and foraging conditions.

HONEY BEE SWARMING

Swarming is a natural process in the life of a honey bee colony. Swarming occurs when a large group of honey bees leaves an established colony and flies off to establish a new colony, essentially creating two from one. Swarming is a natural method of propagation that occurs in response to crowding within the colony. Swarming usually occurs in late spring and early summer and begins in the warmer hours of the day.

Honey bee swarms may contain several hundred to several thousand worker bees, a few drones and one queen. Swarming bees fly around briefly and then cluster on a tree limb, shrub or other object. Clusters usually remain stationary for an hour to a few days, depending on weather and the time needed to find a new nest site by scouting bees. When a suitable location for the new colony, such as a hollow tree, is found the cluster breaks up and flies to it.

Honey bee swarms are not highly dangerous under most circumstances.Swarming honey bees feed prior to swarming, reducing their ability to sting. Further, bees away from the vicinity of their nest (offspring and food stores) are less defensive and are unlikely to sting unless provoked.

In most situations when a honey bee swarm is found on a tree, shrub or house you do not need to do anything. Swarms are temporary and the bees will move on if you patiently ignore them. Stay back and keep others away from the swarm, but feel free to admire and appreciate the bees from a safe distance.

Only if a serious health threat is present because of the location of the swarm, such as in a highly traveled public area, should you attempt to do anything with a swarm. Only as a last resort, you can spray a swarm of bees with soapy water (up to 1 cup of liquid dishwashing detergent in a gallon of water) to kill the bees. Spraying a honey bee swarm is a risky operation because of the large number of bees.

COLD TEMPERATURE AND HONEY BEES

Winter HiveHoney bees will keep the inside temperature of the cluster at about 95 degrees by shivering and expending energy. A bee does not die until its body temperature is 41 degrees. At 41 degrees the bee can't operate or flex it's shivering muscles to stay warm.

The honey bee can survive cold temperature better than wind. It is the wind that is most harmful. Windchill is the single most harmful factor that results in bee deaths in cold climates. If a beehive is located in a shelter or a suitable wind block is supplied, the hives have the best chance for survival.

Last year, the windchill rather than cold temperature correlated significantly with honey bee losses. Windchill is not considered by USDA as a factor when determining honey bee losses. According to USDA, a claim for honey bee losses is not paid unless the temperature does not rise above 14 degrees for 7 consecutive days. On the Island, we should see 20 degrees by Sunday. We have sustained 6 consecutive days where the temperature has not been above 14 degrees, so a claim for losses will not be covered so far this year.

Our location suffered the coldest winter in 70 years but unfortunately, our 2013/2014 losses were not covered by the Emergency Livestock Assistance Program, (ELAP) since our location experienced only 5 days that met the criteria. No losses were covered because most of the winter the cold vacillated between negative temperatures and just above 14 degrees.

Outside the Hive

Often times a honey bee, on its flight back from collecting nectar or pollen, will stop along the way. This is not because the bee is tired and needs a rest. This is the result of flying at 12-15 miles per hour with the wind velocity cooling the bee. The bee stops to flex its flight muscles to warm up.

Bees are cold blooded and there is a limit to the amount of heat they can generate by metabolizing honey and exercising their wing muscles. Clearly, they can stay warm enough when active at outside temperatures in the low 40's, but the warmth will dissipate fairly quickly at colder temperatures and as activity decreases.

I frequently see my Russian bees in flight and gathering nectar on 40 degree days whereas Italians and Carnolians require 50-57 degree days. This is why Russians are cold hardy.

HOW LONG DO HONEY BEES LIVE?

The life span of honey bees varies, depending on the role of an individual bee within the colony. Honey bees live in sophisticated, well organised colonies - they are 'superorganisms'. A honey bee colony could comprise between 50,000 and 60,000 bees, performing different roles in order to help ensure the smooth running and success of the colony.

The role of a bee, as well as the time of year in which it was born (spring/summer or autumn), is also a factor in the lifespan for worker bees in the colony. Worker bees born in spring/summer have shorter, busier lives, whilst those born in autumn may live longer, but they must survive the harsher winter conditions to emerge from the hive the following spring.

In summary:

Queen Honey Bees

All being well, a honey bee queen could live for 3 to 4 years, as long as she is free from disease. It can even be longer - some beekeepers report queen survival of up to 6 years. This is much longer than bumble bee queens or the solitary bee species.

However, a queen that is no longer favoured by the colony may be removed by the workers. In such a case, a new queen will be produced, and the old queen replaced. This is called 'supersedure'. In some beekeeping practices, the queen is replaced by the beekeeper after one or two years.

The bumblebee queen is the founder of new bumblebee colonies.....

The bumblebee queen will emerge from hibernation, already impregnated from the previous year, and will then begin to feed. Nectar will provide her with much needed energy, whilst pollen will give protein, and help her ovaries to develop.

She will then begin the important endeavour of finding a suitable nest site. You may see them early in the year, zig-zagging low across the grass, exploring banks and hedgerow bottoms, and possibly bumping into windows as they try to find a suitable place to establish a nest and rear a colony.

Nest sites, due to habitat destruction, have become increasingly scarce, and it is believed queens will even fight over suitable sites.

Once a suitable nest site has been found, the queen will then begin to establish her colony. Seeing a queen bumblebee carrying pollen loads on her legs is a signal that the establishment of a colony is now underway.

The queen bumblebee is larger than the workers and males she produces. Unlike the female workers, she has the ability to control the sex of the eggs she lays through the use of a chemical signal (pheromone). Worker bumblebees can produce eggs too, but their offspring will be male bees only.

Queens will sometimes face competition from cuckoo bumblebees. Cuckoos attempt to take over a nest, sometimes (but not always) killing the resident queen if successful.

Unlike the host, cuckoo species are unable to carry pollen and secrete wax to make cells.

At the end of the season, only the newly emerged queen bumblebees will survive (although there are some exceptions in warm climates). They will mate, and hibernate to establish new colonies the following year. New queens emerge from colonies toward the end of the season, and begin feeding to lay down fat reserves ready for hibernation. In truth, a colony can only be deemed successful if it has had the opportunity to successfully produce queens that can mate and hibernate, and thus ensure future generations.

Quick Facts About The Honey Bee Queen

The honey bee queen is the largest of the bees in a honey bee (Apis mellifera) colony, measuring around 2 cm - that's about twice the length of a worker - drones are slightly larger than workers.

For the human eye, despite being larger than the workers, honey bee queens are difficult to spot among thousands and thousands of worker bees. For this reason, beekeepers mark queens with a dot of special paint on the thorax, as can be seen in the photograph above.

As she establishes her colony, she may lay 1000 eggs per day – one egg every 20 seconds - and more than her own body weight in eggs!

If a queen is not performing very well (for example, if she is not laying enough eggs), the colony may decide to replace her with a new queen. This is called supersedure.

In a colony of 50,000 bees, there will be only 1 queen, and perhaps around 300 drones (males) and the rest will be female worker honey bees. At exceptional times, there may temporarily be 2 queens, but not for very long. More about this below.

HOW ARE QUEEN HONEY BEES DIFFERENT FROM WORKERS?

Apart from being significantly larger than workers, the roles that queens perform is very different from that of other members of the colony.

The role of the queen honey bee is:

to mate with drone bees (males),

produce eggs,

and to begin new colonies through swarming.

This is in contrast to the role of workers. In an active colony of honey bees, depending on age, workers perform a range of activities.

The role of worker honey bees includes:

foraging for food

building honey comb, which includes secreting wax from their abdomens

keeping the nest clean

defending the nest from predators

feeding the larvae

The main role of drones is to mate with honey bee queens and thus ensure future generations of honey bees. They also help to control the temperature in the hive or nest .

How Do Honey Bees Become Queens?

Despite the difference between herself and the female workers, what may surprise many is that she is produced from eggs that are in every way identical to those eggs producing workers. The difference, however, is that larvae of potential queens are fed only a special substance called 'royal jelly'.

Mating Behaviours

About a week after a new queen emerges from her cell, she will take several flights in order to mate. She may mate with as many as 20 drones, all while in the air! (The drones, unfortunately, die after mating). However, when the honey bee queen returns to lay her eggs, she will only rarely leave the colony after that.

Inside her, she will have enough sperm (which she stores in her sperm pouch or spermatheca), so that she may continue to fertilise her eggs for the rest of her life.

When she returns to the colony from her nuptial flight, and now impregnated, the workers begin fussing over her.

They feed her so that her abdomen swells, and lick her – a process which transfers a chemical (pheromone), used to regulate the colony.

The Queen Pheromone And Communication

Pheromones are produced by the drones and workers, as well as the queen, who produces a 'queen pheromone'.

The queen pheromone encourages workers to tend to her and the brood, whilst at the same time, inhibits the production of more queens.

So efficient is the pheromone for communicating within the colony, that if the queen is removed from a hive, within 15 minutes, all of the bees will know about it, and will frantically begin the task of creating a replacement!

When colonies become very large so that workers cannot (due to the distance between themselves and the queen), detect the queen pheromone, then this encourages part of the colony to create a new queen. A new colony will then be formed.

This initiates 'bee swarming'. A clump of workers surrounding a queen honey bee might be seen resting temporarily on a tree branch or post whilst 'scout bees' are looking for a suitable place for a permanent nest.

Workers

Workers raised during the spring or summer months may live for 6 or 7 weeks. Their lives are especially busy, with lots of hungry larvae to feed, and honeycomb to be produced. This is when the colony is at its most productive, with workers busy collecting nectar and pollen for feeding the colony.

Workers raised in the autumn have no brood to care for, since the queen stops producing eggs. These workers, together with the queen, comprise the remains of the colony for the year (part of the colony may have left in a swarm, in order to form a new colony elsewhere), and they must huddle together around the queen in order to keep warm during the winter, ready to emerge the following year to begin foraging again in the early spring. They may live 4 to 6 months.

Drones

At the most, drones may live for up to 4 months, however, they may survive for just a few weeks. Note, that upon mating with the queen, drones die immediately.

Drones are fertile male honey bees, and they are vital for the survival of honey bee colonies. Their primary role is to mate with a receptive queen honey bee, in order to ensure future generations of honey bees, and indeed, expansion and creation of new colonies.

I've heard it said that drones don't do much because (say some):

Drones do little around the hive (or, if a feral bee perhaps a hollow tree or cavity in an old building somewhere) - they don't clean or build honey combs, for example.

They help themselves to nectar stores, yet they don't do much to help out with the kids (okay, okay, I'm humanising them a bit – the brood!).

Heck, they don't even go out and get food for the colony!

It is sometimes said that drones spend their time drinking nectar, mating (in the air at that!), and lazing around on flowers.

Infact, one scientific paper by Kova et al remarks:

"Honeybee drones are often called "lazy Willi" (Bonsels, 1912) and are often assumed to merely function as "flying sperm", necessary to inseminate virgin queens".

The scientist goes on to say,

"This view, however, is not correct.".

The fact is:

Drones perform precisely the role that nature gave them.

Below, you can read about the importance of honey bee drones.

- They may live for just a few weeks or up to 4 months.
- They die straight after mating!
- Not all drones get to mate.
- They cannot sting.

- At the end of the summer, or when the going gets tough, they're the first to be kicked out of the colony, so as not to drain resources.
- Drones are fatherless.....yet they have a grandfather!
- It takes 24 days for the drone to develop from being an egg to a fully grown adult bee.
- Drones are essential to the health and survival of future honey bee colonies.

How Large Are Drone Bees?

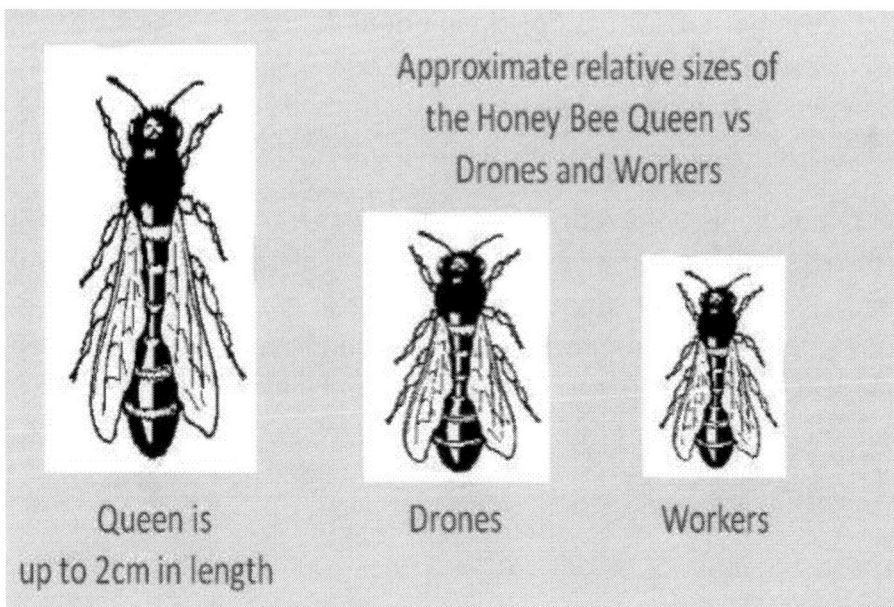

Approximate relative sizes of the Honey Bee Queen vs Drones and Workers

Queen is up to 2cm in length

Drones

Workers

- Drones can vary widely in size, but they are larger than workers, and smaller than queen honey bees. The cells they develop from are slightly larger than worker cells.
- They also have significantly larger eyes in comparison with queen and worker honey bees.

Role Of Drones In the Honey Bee (Apis Mellifera) colony

Drones ensure the continuation of honey bees as a species, by mating with queens.

Drones can pass on important behavioural traits to new generations of honey bees, (such as hygenic behaviours) through their genes.

Drones help to regulate the temperature in the hive or nest, and this is especially important for the development of young bees and larvae. Honeybee larvae and pupae are extremely stenothermic, which means they strongly depend on accurate regulation of brood nest temperature for proper development (33–36°C).

Although each colony has far fewer drones than workers, they nevertheless pull their weight with regard to heat generation. Research indicates that Drones can produce one and a half times as much heat as a worker bee, and that even those Drones not directly next to the brood, are never the less assisting with heat regulation inside the nest.

Fatherless Drones

Drones are 'haploid', having been reared from an unfertilised egg. As a result, a drone has only half the chromosomes of a worker bee or queen bee the drone has 16 chromosomes, workers and queens have 32. (That honey bee drones are haploid was first discovered in 1845 by a Polish apiarist, named Jan Dzierżon often described as the "father of modern apiculture".

eing haploid means that drones can have a grandfather and grandsons, but a drone cannot have sons!

To explain further:

Drones come from unfertilized eggs (they are 'haploid'), meaning that no male (drone) was needed in order for the queen to produce more males in other words, they are formed without a male 'parent'. (In fact, at various times, female worker bees can and do also create drones, despite the fact that they do not mate with males - they are referred to as 'Drone Laying Workers').

But what about the queens - the mother of all the drones?

In order for a queen to produce new queens, eggs must be fertilized, for which drone bees are necessary. This means that the queen (the "mother" of the drone), has a "father" - obviously a drone. This means that any drones produced by the queen, actually have a "grandfather" (i.e. the "father" of the queen) more specifically, a "maternal grandfather" (the father of the queen).

Put yet another way and more succinctly:

The queen who laid the drone eggs, is the offspring of an egg fertilized by a drone (male). Drones themselves, however, are the offspring of eggs that have not been fertilized by a male, and they are therefore, fatherless.

This scenario, whereby offspring are reared from unfertilized eggs, is referred to by biologists as 'parthenogenesis'.

Drone Mating Behaviour

Each honey bee colony will produce several hundred drones (in contrast with the thousands of workers).

On warm and sunny afternoons during the mating season, sexually mature drones, fly out of the nest (or hive) and congregate with other drones high in the air, to form a cloud of bees. There may be as many as 11000 drones from up to 240 different colonies.

These clouds of drones can measure between 30 and 200m in diameter, and be located 10–40 m above ground.

About one hour after the peak of drones' departure from the hives, virgin queen will also leave her hive for her nuptial flight, and join the drone congregation.

As soon as a virgin queen enters the congregation of drones, groups of drones are attracted to her, first by olfactory cues (pheromones), and at shorter range by visual cues. Drones follow the virgin queen in a comet-like swarm each competing to approach and mate with the queen.

Usually, a queen mates within 15–30 minutes, and with just 10–20 of the thousands of drones, and each drone that mates with the queen will die after mating .

This happens because the drone's reproductive organs are torn away from its body, whilst the queen flies off, with the drones genitalia attached to her.

How Long Do Drone Bees Live?

Drones may live for just a few short weeks, but it is also possible they may live 3 to 4 months.

They are expelled from their colonies by the end of summer, but in any case, by the end of autumn, there will be few or no drone bees around.

Queen bumble bees

All in all, providing a bumble bee queen is successful and is not harmed by disease, pesticide or predators, she may live for about a year - part of this time being spent in hibernation.

New queens emerge during the late summer or early autumn. After mating and feeding to store fat reserves on their bodies for the winter, they will then hibernate. During early spring the following year, the new queens emerge to establish new colonies of their own.

The bumblebee queen is the founder of new bumblebee colonies.....

The bumblebee queen will emerge from hibernation, already impregnated from the previous year, and will then begin to feed. Nectar will provide her with much needed energy, whilst pollen will give protein, and help her ovaries to develop.

She will then begin the important endeavour of finding a suitable nest site. You may see them early in the year, zig-zagging low across the grass, exploring banks and hedgerow bottoms, and possibly bumping into windows as they try to find a suitable place to establish a nest and rear a colony.

Nest sites, due to habitat destruction, have become increasingly scarce, and it is believed queens will even fight over suitable sites.

Once a suitable nest site has been found, the queen will then begin to establish her colony. Seeing a queen bumblebee carrying pollen loads on her legs is a signal that the establishment of a colony is now underway.

The queen bumblebee is larger than the workers and males she produces. Unlike the female workers, she has the ability to control the sex of the eggs she lays through the use of a chemical signal (pheromone). Worker bumblebees can produce eggs too, but their offspring will be male bees only.

Queens will sometimes face competition from cuckoo bumblebees. Cuckoos attempt to take over a nest, sometimes (but not always) killing the resident queen if successful.

Unlike the host, cuckoo species are unable to carry pollen and secrete wax to make cells.

At the end of the season, only the newly emerged queen bumblebees will survive (although there are some exceptions in warm climates). They will mate, and hibernate to establish new colonies the following year. New queens emerge from colonies toward the end of the season, and begin feeding to lay down fat reserves ready for hibernation. In truth, a colony can only be deemed successful if it has had the opportunity to successfully produce queens that can mate and hibernate, and thus ensure future generations.

How Long Do Bumble Bees Live?

Information about the lifespan of bumble bees tends to vary between different studies. Additionally, in the tropics, where colonies may thrive for longer, the lifecycles and lifespan of colony members may be quite different.

There are other anomalies - for example, even in the UK, climate is believed to be having an impact on bumble bees, with some colonies continuing to be active during the winter in the south of England. However, typically…

Worker bumble bees

Accounts of how long bumble bees live, do vary between species and studies. For example, Bombus terricola workers were observed to be 13.2 days on average – around 2 weeks, but in other studies, workers were observed to live for up to 41.3 days – about 6 weeks.

It is believed that workers engaged in nest duties live longer than bumble bee workers whose main duty is foraging. These bumble bees are of course more prone to predator attack, and are also exposed to varying weather conditions. It is not unusual to find a bedraggled looking bumble bee needing to rest and revive itself, having been caught in a shower. For advice about what to do if you find such a bumble bee, read my page advising what to do if you have found a bee.

The Cuckoo Bumblebee

Cuckoo bumble bees belong to the sub-genus, Psithyrus. They are 'social parasites' and are so called, because like the bird, this species lays its eggs in the nests of true social Bombus species, to be reared and fed by the host.

The tendency of some species to become nest parasites of others is sometimes called 'inquilinism'.

There are 10 species of cuckoo bumble bee in Europe – 6 species may be found in Britain (the image above of Bombus barbutellus is found in Britain.

According to The Bees in Your Backyard: A Guide to North America's Bees there are 29 species of Bombus (Psithyrus), 6 of which occur in the United States and Canada.

You can read about the status of cuckoo bumble bees found in North America (according to the IUCN) here.

How Are Cuckoos Different From Social Bumble Bees?

The cuckoo has evolved a number of characteristics, which mean it is totally reliant upon its host for its future survival.

Firstly, cuckoo species are unable to establish their own nests. They do not have the ability to excrete wax from their abdomens for making egg cells in which to lay their eggs, nor can they make honey pots from which newly emerged brood may feed, and from which they may feed themselves whilst they incubate the brood (see the bumble bee life cycle). The cuckoo female must use the larval cells and cups made by the host queen.

The cuckoo female also cannot collect pollen for returning to her nest, since she has no pollen baskets (or corbicula) on the hind legs, although they do of course, eat pollen, like social bumble bees do. Pollen is important as it helps the ovaries to mature in the fertilized females, and it is used for feeding the brood.

Also, cuckoos are unable to rear workers! The cuckoo merely lays females and males like herself, and these offspring are likewise limited, in that they cannot perform the tasks that worker bees of the social species can undertake, in order to sustain the colony (building wax cells and collecting pollen).

So How Do Cuckoo Bumble Bees Take Over A Host Nest?

Cuckoo females typically emerge from hibernation a few (perhaps around 6) weeks after the target host species. Some cuckoos target only one specific species of bumble bee, whilst others may select from two or three target hosts.

It must firstly locate a ready prepared nest at an appropriate stage of development.

Of course, bumble bee workers are very important for increasing the size of a colony and rearing the brood. For this reason, the female cuckoo must choose her target host carefully. There must be a colony already established of at least a few workers to help the cuckoo rear her offspring. On the other hand, if there are too many workers in the nest, she may be attacked and easily defeated.

A nest with 2 broods already reared, may have enough workers to overpower the cuckoo, and prevent her establishing herself in the nest.

Once the cuckoo has found a suitable nest, she will typically lurk around it for some time, in order to 'pick up the scent' of the host.

This will help her to usurp the resident queen if necessary, and gain acceptance of the workers through physical attacks and the use of her pheromones.

Once the cuckoo enters the nest, there are a variety of scenarios that may occur, but a common one is that the original host queen is killed.

The cuckoo will then set about laying her own eggs in the nest for the workers of the original queen to tend to and feed.

As stated, however, timing is critical. If the cuckoo has entered a nest that is well developed, and there are many workers, they may attack the parasite, and kill her. On the other hand, if there are too few workers to support her, then she will not be able to rear many offspring.

Is the true queen always killed?

Very often, she will be.   Although cuckoos closely resemble social species, the female parasite is often a little larger than her host. She has a more powerful, longer sting, and a thicker, though less hairy coat. She can often overpower the founder queen unless there are sufficient workers to attack her.

However, there have been reports of both host and cuckoo cohabiting in a nest for quite some time, with the cuckoo even incubating the host queen eggs, and offspring from both the cuckoo and host queen emerging!

For example:

A study by Hoffer in 1889 observed that offspring were reared from both the cuckoo and the host queen, when the cuckoo Bombus campestris targeted Bombus pascuorum as its host; and when Bombu sylvestris, another cuckoo species, layed its eggs in the nest of Bombus pratorum.

Studies more recently, have also found that the outcome can vary significantly.

Küpper & Schwammberger in 1995, studied Bombus pratorum nests that were invaded by Bombus sylvestris. Again they found that both species reared offspring from the same nest!

Research has also shown that the cuckoo is not always successful at establishing herself in the host nest. She may be attacked and killed by the target host queen and workers.

In addition, it has been observed that some cuckoos may take temporary refuge in the nests of non-target host species, without attempting to usurp the queen.

British Cuckoo Bumble Bee Species And Their Hosts

There are some that typically select just one species as its host, whilst others target more than one type of social bumblebee.

For example, of the UK species:

Bombus bohemicus - Gypsy Cuckoo Bee

targets Bombus lucorum.

Bombus vestalis -Vestal Cuckoo Bee

targets Bombus terrestris.

Bombus barbutellus - Barbut's Cuckoo Bee

targets both Bombus hortorum and Bombus rudertatus.

Bombus rupestris - Red-tailed Cuckoo Bee

targets Bombus lapidarius.

Bombus campestris - Field Cuckoo Bee

targets Bombus pascuorum, Bombus humilis and previously, Bombus pomorum (now extinct in Britain, but likely due to habitat loss).

Bombus sylvestris - Forest Cuckoo Bee

targets Bombus pratorum and possibly Bombus jonellus and Bombus monticola.

North American Cuckoo Bumble Bees And Their Hosts

There are 6 Bombus pysithyrus in North America:

Bombus suckleyi - Suckley Cuckoo Bumble Bee - recorded as breeding as a parasite of colonies of Bombus occidentalis.

Also recorded as present in the colonies of Bombus terricola, Bombus rufucinctus, Bombus fervidus, Bombus nevadensis and Bombus appositus.

Bombus suckleyi - Suckley Cuckoo Bumble Bee - recorded as breeding as a parasite of colonies of Bombus occidentalis.

Also recorded as present in the colonies of Bombus terricola, Bombus rufucinctus, Bombus fervidus, Bombus nevadensis and Bombus appositus.

Bombus insularis - Indiscriminate Cuckoo Bumble Bee -

recorded as breeding as a parasite of colonies of

Bombus appositus, Bombus nevadensis, Bombus fervidus, Bombus flavirons, and Bombus ternarius.

Also recorded as present in the colonies of Bombus terricola, Bombus rufucinctus, Bombus occidentalis, and Bombus nevadensis.

Bombus bohemicus - Ashton's Cuckoo Bumble bee - recorded as breeding as a parasite of colonies of Bombus terricola and Bombus affinis.

Also likely to breed in colonies of Bombus occidentalis and Bombus cryptarum.

Bombus variabilis - Variable Cuckoo Bumble bee -

recorded as breeding as a parasite of colonies of Bombus pensylvanicus.

Bombus citrinus - Lemon Cuckoo Bumble Bee -

recorded as breeding as a parasite of colonies of Bombus impatiens, Bombus bimaculatus, and Bombus vagans.

Bombus flavidus - Fernald Cuckoo Bumble Bee - No direct records of it actually breeding in the colonies of host species in North America, though it has been recorded as present in colonies of Bombus rufucinctus, Bombus occidentalis and Bombus appositus.

HOW LONG DO BEES LIVE?' AND THE LIFE CYCLE OF BEES

The bee life cycle has 4 key stages, but even the length of time between those stages varies, depending on species and role within the colony. To read more, see may page about the bee life cycle.

Some bees hedge their bets!

According to book, The Bees In Your Backyard, some bee larvae wait in their nests before emerging, with only some of the larvae developing and emerging from their nests the following year - and this is called 'bet-hedging'.

For example, not all larvae of the Habroda species emerge from the nest in the same year! Whilst some will emerge, others may remain in the nest for up to 2 years. However, the book also points out that some will remain in the nest, but delay emergence for 7 or even 10 years!

It's thought that hedge-betting behaviour in bees is especially beneficial in challenging conditions such as the desert. If only some of the bees emerge the following year (whilst others remained behind), then not all of the bees will perish if there are insufficient flowers upon which to forage.

This is the case with the desert bee, Perdita portalis. In a study by Cornell University: "Emergence dynamics and bet hedging in a desert bee, Perdita portalis":".....according to new research by Cornell entomologist Bryan N. Danforth, not all the viable larvae emerge in any one year of diapause, and their "coming out" is triggered by rain. "

Writing in the October issue of The Proceedings of the Royal Society of London, Danforth notes that by spreading reproduction over several years, desert bees can keep catastrophic losses of their kin to a minimum in very bad drought years."

<u>WHAT TO DO TO REVIVE BEE</u>

Please note that if the weather is dry, and the bee is simply motionless on a flower, leave it alone. It is simply resting (scientists have confirmed that bees sometimes exhibit characteristics which could be described as 'sleep'), and will become active again in its own time. Don't be tempted to interfere in such cases.

In general, it is best to allow the bee to remain outside, where it can get access to flowers. However, if it's raining, and you have found a bee that is very wet and can hardly move, if possible, bring it into a dry place. If you place the bee in a box, ensure there are sufficient air holes so that it can breathe. Please note, your aim must be to release it again as soon as possible preferably immediately if the bee feeds straight away see below.

If I have found a bee should I feed it?

You may offer the bee a solution of sugar water made with clean water and ordinary granulated sugar (do NOT use artificial or diet sweeteners or demerera sugar). The ratio of sugar to water should be about 1:2 (i.e. one spoon of sugar to two spoons of water).

Do not use honey, because honey may contain traces of viruses that may be passed on to the wild bee.

Even if you are attempting to revive a honey bee, do not feed it honey honey bees should only ever be given their own honey, and should not be given honey from other colonies, even if it is organic.

The best way to offer the sugar water is to sprinkle drops of water on to a clean, solid work surface near to the bee, or dip a flower head into it - this could be a dandelion head, for instance. One gentleman wrote to me to say that he uses moss.

If this is not possible, use a spoon. It's important to ensure the bee will not fall in to the liquid and drown (if there is a lot of liquid).

If it is dark, you could keep the bee indoors. Find a place that is not too hot or cold, where it can remain undisturbed over night.

As stated, the aim is to set the bee free as early as possible in the morning.

You could offer more sugar water first, but it's best to let it go and find some flowers.

Please note, bees, and especially wild bees, do not make pets! (Yes, I have been asked). For one thing, bees need to get back to their colonies as quickly as possible. Secondly, sugar water is not a suitable alternative for nectar and pollen for very long, since pollen and nectar contain other nutrients and fats needed by bees which are not present in sugar. Bees need to be able to forage in the outdoors to gather what they need, and take it back to their colonies.

The bee may even be a queen, in which case, queens establish the next generation of bees, and they need to be allowed every possibility to do just this.

Also, bees have short life cycles, in which the busy bee has much to do in a matter of weeks! New queen bumble bees need to have the opportunity to mate, feast on pollen, then find a place to hibernate, and they need to be free and outdoors to do this.

HONEY BEE LIFE CYCLE

The honey bee life cycle goes through 4 basic stages. More detail is added in the diagram below, but the key stages are:

Egg

Larva

Pupa

Adult

All bee life cycles go through these stages, although there are great variations between the life cycles of solitary, honey and bumblebees.

Larva spins a cocoon, and pupates.

Not much room in here!

Larva reaches its full size. Brood nurses seal the egg cell.

Adult honey bee emerges from cell.

Hello!

Workers (brood nurses) feed the larva 1,300 times a day!

phew!

When I grow up I want to be... a queen of a drone.

Potential queens are fed royal jelly.

Egg is laid by the Queen in an egg cell. The egg is about half the size of a grain of rice.

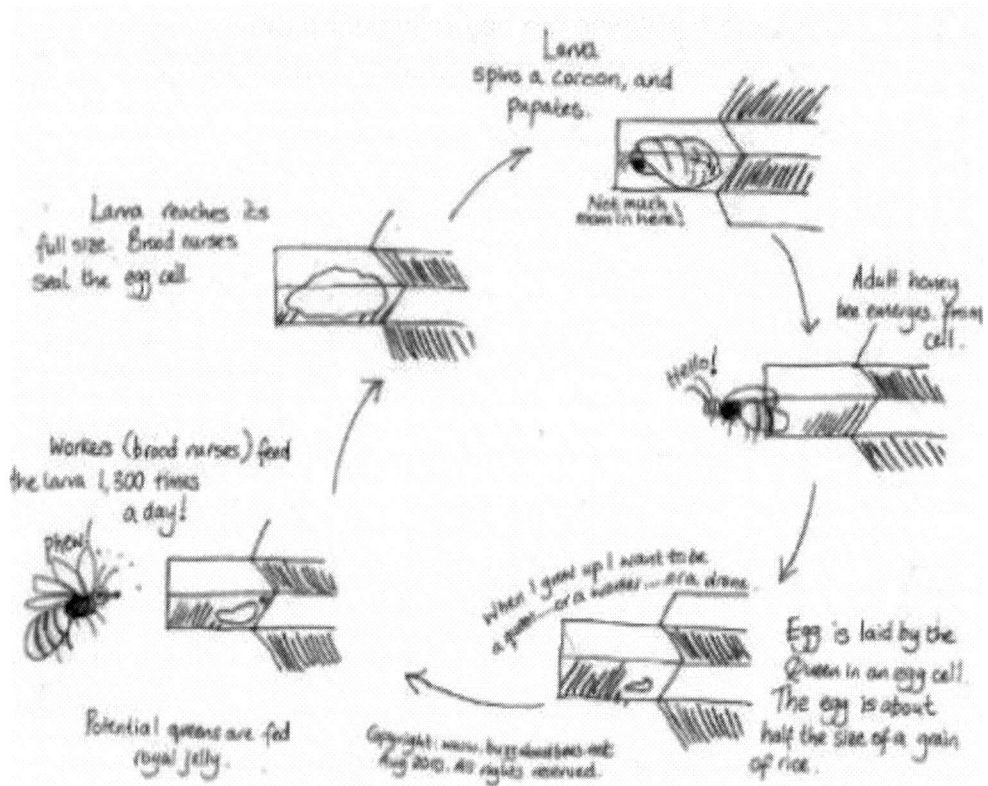

bumblebee colonies, honey bee (Apis mellifera) colonies can survive the winter, provided they have enough food resources, are able to keep sufficiently warm, and are free of diseases and predators. However, in the winter, colonies are smaller than in the summer: there are no drones, and perhaps part of the colony left the hive (in a swarm) to form a new nest elsewhere.

Some of the workers will also die naturally during the winter months, however, there may be up to 20,000 workers left, and a queen.

The queen and the rest of the colony will form a winter cluster to keep warm during the cold months. There will be no brood to tend to, and no eggs are laid during this time. However, as the days begin to warm up, and the flowers begin to bloom, honey bees will begin to go out foraging again, and the queen honey bee will begin to lay eggs.

After 3 days, eggs hatch into worker larvae. During this stage, each larva will be fed about 1,300 times a day! They are fed by worker bees that have the specific task of tending the brood, and are referred to as the 'brood nurses'.

The food given is made from pollen, honey, and secretions from the brood nurses, and is called 'bee bread'. (Find out more about bee bread).

Potential honey bee queens, however, are given 'royal jelly', a much richer food.

After about 6 days, the egg cells are capped, and each larva spins itself a cocoon and becomes a pupa.

Worker bees take 10 days to emerge from pupae. Drones take slightly longer. New Queens, however, take about 6 days.

Drones:

Drones may live just a few weeks, or they could live up to 4 months. Drones that mate with new honey bee queens, will die immediately after mating.

By the end of the summer, they will no longer be needed by the colony. Honey bees need reasonable weather to forage, and of course, during the winter time, there is far less nectar and pollen available.

Drones do not collect pollen or nectar, and those still alive will be elbowed out by the workers, so that winter food resources are not drained! Learn more about drones.

Workers:

Workers raised in the spring and summer have shorter, busier lives, and may live 6 or 7 weeks. This is the most productive time for the colony, with larvae to be fed, nectar and pollen to be gathered, and honeycomb to be built.

Those raised in the autumn will have far less to do, with no brood to care for. Their main concern will be to survive the cold until the following spring. However, they may live 4 to 6 months.

Whereas the queen honey bee life cycle revolves primarily around mating and laying eggs, the life of worker honey bees progresses through various stages of functions within the colony.

Queen Honey Bees:

A productive queen, favoured by the colony and free from disease should certainly live for about 2 yrs, but could live for up to 3 or 4 years, partly depending on whether the beekeeper decides to get rid of the queen, or whether the colony decide to replace her. The act of deposing the queen by the colony is called 'supersedure'. Learn more about the role of the Honey Bee Queen.

The diagram below give you an idea how large the queen is, in comparison with the workers and drones.

Approximate relative sizes of
the Honey Bee Queen vs
Drones and Workers

Queen is
up to 2cm in length

Drones

Workers

Queens are difficult for beekeepers to identify among the thousands of workers, and so they are marked with a special type of paint, as can be seen by the dot of white paint on the photograph below

So this has given you a brief summary of the honey bee life cycle, but you can learn a lot more about some of the specific stages, such as swarming, by reading further.

URBAN BEEKEEPING

Urban beekeeping has been on the rise, probably for a combination of reasons. People have ventured into beekeeping because they have heard about bee decline, and wanted to do something about it. This has, on some occasions, extended to businesses and even churches. In London, for example, St Paul's Cathedral and the Museum of London have become home to several colonies of honey bees, with hives installed on the roofs. On the other hand, some people have an interest in producing honey and others have simply decided to pursue a hobby.

Whatever the reason, due to the fact that so many people live in towns and cities, its urban beekeeping or no beekeeping for many. However, whilst bees can thrive in towns and cities, urban beekeepers do need to take into account several issues particular to their environment, so if this is something you are contemplating, maybe you would like to think about the following points.

Urban Beekeeping And Considerations For Urban Beekeepers

Temperament of the bees

The temperament of the bees you install in your hive of bee hives, is something you should think about. It is better to avoid more aggressive bees, and to opt for more passive strains. See types of honey bees.

Regional authority restrictions

Is urban beekeeping allowed in the area in which you live? For example, in some towns in the USA, there are restrictions on beekeeping. If you have a city allotment, you may have to negotiate with the body in charge of the allotment (and even the other allotment holders).

Floral provision for bees

Which leads to the next point: what kind of provision is there for the bees, in terms of flora? Are there plenty of private gardens with an abundance of nectar and pollen-rich flowers? Are there sufficient flowering trees and hedgerow? Parks and public planting schemes can provide opportunities, but caution is advised if the local council uses insecticides such as neonicotinoid treatments. Quality is important, not merely quantity. Rowsof highly cultivated, brightly coloured annual flowers will be of little benefit.

Traditional single-petal flowers, herbs, and nectar-rich flowering shrubs and trees are better (such as hawthorne, berberis etc). They'll need food from early spring to autumn. Also, don't forget about your own contribution: if all you have is a balcony, do your bit with hanging baskets, window boxes and planters.Wildflowers on brownfield sites and waste areas may also provide food, but if you live in an area of rapid building development, some of these may disappear.

Vandalism

Potentially, vandalism could represent a barrier to urban beekeeping.Are vandals likely to be a problem in your area? Unfortunately, if you live in an area with high levels of vandalism, it may be unwise to install hives, although some urban community projects successfully maintain and protect them. Close to where I live, I heard of a proposal to install beehives in an area in an attempt to deter vandals from destroying ancient ruins. The plan was quickly abandoned when the inexperienced (albeit well-meaning) individuals realised there is nothing tostop vandals throwing bricks at hives from a distance, for example. ……Although I understand bee hives with colonies of honey bees are very effective for discouraging elephant crop raids in Africa! However, irresponsible vandals are a different matter.

Beekeeping support

Is there a local and active beekeeping association in the area? If you are new to beekeeping, this may be very important for you.They would also be able to share advice and tips.However, if you are venturing into urban beekeeping as a means to 'do your bit' for the bees, then speak with local beekeepers first – and perhaps even your local conservation body. In 2010, the Mayor of London, England came under criticism from the local beekeeping organisations, for encouraging more beekeeping (see below): "John Chapple, chair of the London Beekeepers' Association, which has seen a five-fold increase to 150 members in the past few years, said: "London is already saturated with beekeepers. We don't need any more, what we need are better beekeepers.....Rather than jumping on the beekeeping bandwagon, Boris [the Mayor] should stop parks from planting double-headed flowers that provide no nectar or pollen, cutting back trees and shrubs that provide vital forage for bees, and spraying with chemicals".

Your neighbours

What is the attitude of your neighbours? If they have young children (or even if they don't), they may be against you keeping bees for fear of being stung.

You may be able to allay their fears, and take measures to discourage the bees from swarming, and you may be able to erect a tall screen to encourage the bees to 'fly high'.

However, if you are not able to come to some agreement with your neighbours, it may be better to find an alternative location for your hive perhaps at an allotment (if you are permitted) or the garden of a friend with neighbours who are amenable to the idea of living next door to a honey bee colony. Note that some people do have very legitimate reasons for concern allergy to bee stings is not common but can be fatal! It is also the case that some neighbours can merely be awkward, mean or jealous. Depending on where you live, it may be worth you checking what your rights and responsibilities are before investing your time, money and energy into keeping bees. This is another reason why linking up with other beekeepers in the area who are both knowledgeable and experienced, is so helpful.

Pollution

Pollution is another important issue to consider in urban beekeeping. If chemicals are used in the local park next door (sprays, lawn treatments, and even insecticides applied to the soil), it may put your bees at risk. Also, think again about urban beekeeping if all you can provide is a tiny front yard facing a very busy road with the bees on the same level as all the petrol and diesel fumes, or next door to a factory churning out noxious fumes. In such a circumstance, again, you may be better off finding an alternative location to keep your bees. A study by researchers from Pennsylvania State University found that air pollutants alter floral scent molecules.

Pollutants such as ozone, nitrate radical and hydroxyl radical are responsible for "the degradation and the chemical modification of scent plumes."Bees pick up scent molecules from flowers at distance, thus enabling them to locate sources of food, and thus pollinate flowers. Unfortunately, researchers found that the pollutants have the effect of reducing the distance scent molecules can spread. This means it becomes harder for bees to locate flowers, with the knock on effect that foraging times are increased.

Furthermore, the study showed that even a moderate level of air pollutant at 60 parts per billion of ozone, can make it harder for bees to locate flowers. This alteration could have a dramatic effect on bees' foraging habits and food supplies that could create "severe cascading and pernicious impacts on the fitness of foraging insects by reducing the time devoted to other necessary tasks," the study's abstract suggests.

But what levels of pollution are we talking about here? Actually, even 'moderate' levels have this effect - 60 parts per billion and upwards. With increasing levels of pollution, there potentially more danger to bees.

Are roof top bees the answer?

A roof top can be an ideal place to put a hive, but remember the hive will need protection from wind, and you will need good, safe access to the hive. The hive will need to be stable, whatever the whether, and you'll need good access.

WHEN HARVESTING HONEY-TIPS & TRICKS

For beekeepers, when it's not science fiction to harvest honey, there are a few indicators to know when to collect honey from the beehive. Generally speaking, the beekeepers harvest their honey at the end of a substantial nectar flow and when the beehive is filled with cured and capped honey. Conditions and circumstances vary a great deal across the country. First-year beekeepers are lucky if they get a small honey harvest in late summer. That's because the new colony needs a full season to build up a large enough population to collect a surplus of honey. We collected the best tips and tricks to harvest honey from a beehive.

Check the hive frequently When the summer season is approaching and you're ready to harvest the honey, it's a good idea to check under the hive cover every two or three weeks. This will allow you to check the progress made, and let you see how many frames are full of capped honey. It's not a good idea to harvest honey at this point; it's best to wait for the end of summer if possible.

If the frame contains at least 80% of the sealed and capped honey, this is when you can start harvesting the honey for the season. If you are a little more patient and want the best results, you can wait a while longer; if you choose this route, there are a few conditions you should be looking for. First, you can wait until all the caps are full of honey; or you can wait until the ultimate significant nectar flow moves, so you can harvest.

Honey can be obtained from open cells (not capped with wax) if it is healed. To see if it's healed, push the cell picture towards the floor. Please give the picture a mild twist. If the honey leaks out of the cells, it's not cured and it shouldn't be extracted. This stuff isn't even cute. It's a nectar that isn't healed. The water content is too high for it to be considered as honey. Attempting to bottle the nectar results in watery syrup, which is likely to ferment and spoil.

You want to wait until the bees have collected all the honey they can, so be patient. Well, that's a virtue. But don't leave the honey supers on the hive too soon! Things tend to get really busy around Labor Day.

In Harvesting Patience Is Crucial

It's best for the harvesters to wait for the bees to gather as much honey as they can. It's better if you can wait for the ultimate stream of nectar, but at the same moment you have to be mindful of waiting too long for the harvest. If necessary, it is best to remove the honey by mid-September at the latest; there are two explanations for this.

First, as the winter season starts to approach, your bees will start to eat the honey they have produced during the summer months. If the supers are kept in the hive for too long, the bees will start to eat the honey they've produced.

The second reason is that when the climate grows too hot, you will no longer be prepared to harvest honey and loose what your bees generated during the summer.

Although it's a sensitive time frame you've got to operate with, you've got to be mindful of getting in too quickly. If you collect eggs before the 80% capped flower mark, you run the danger that birds will no longer produce for the summer. But, you want to harvest before the winter months, to prevent loss as well.

The finest months are likely early July, August and mid-September. Not only will this result in a full frame, it will also guarantee that you do not loose your honey owing to climate circumstances in the previous months.

HOW DO YOU INSPECT THE QUALITY OF HONEY AT HOME?

If you want to appreciate most of the advantages that come from honey. The purity of the baby is what you should consider before purchasing. There are some simple tests and experiments that can be performed at home to check the purity of the honey. Find out what kind of exams you should attempt!

Before carrying out any of the trials, one fundamental and incredibly easy "how to verify the quality of the fruit" technique is to read the tag on the honey bottle prior to buy. Manufacturers are needed to mention additives and extra ingredients that have been added to the honey generated. So you can create out, if organic or additional sweet or artificial flavorings have been added, merely by checking the tag. If you buy honey directly from a beekeeper, then the honey is of pure and unprocessed quality, as you buy it directly from the source. Before doing the trials, we suggest that you know the distinction between false and true honey.

Conducting honey purity exams at home Honey's beautiful, delightful blend works against you when you're attempting to discover a easy exam. Different kinds of plain honey can contain a wide variety of density, flammability and other features. While the following exams are based on real values, your findings may be inconclusive in reality. Try a few of these trials to see if the honey falls or runs continuously. You can get nothing more than a good guess in many cases.

In order to check the purity of honey at home, here's what to do:

Thumb Test

Here's the procedure to do a thumb test:

- Put a small drop of the honey you have on your thumb
- Check to see if it spills or spreads around
- If it does, it is not pure
- Pure honey will stay intact on your thumb

The Water Test to Spot Fake Honey Here's how to do the water test: Fill a glass with water Add one tablespoon of honey to the glass Adulterated or fake honey will dissolve in water and you'll see it around the glass Pure Honey on the other side will lie directly at the bottom of your glass The Flame Test to Know Pure Honey Did you understand that organic honey is flammable? Here's a sample to understand 100% pure organic tea.

Take a dry matchstick Dip its tip straight into the honey Strike the handle on the matchbox as if to raise it If the honey is pure, the matchstick will extend with ease The flame will also maintain the honey flowing. However, if it is impure, it will not be bright because false honey includes moisture as one of the impurities These are some of the easy and popular methods to check pure honey at home. Another popular technique to say the distinction is as follows: add some water and 2-3 layers of vinegar powder to the honey and blend well. If the answer is foamy, that's certainly an adulterated honey.

Drop honey on blotting paper or paper towel If the honey has been diluted with water, it can be absorbed or leave a wet mark on an absorbent material such as blotting paper. Pure honey should not be consumed, but sadly, most sugar syrups are not diluted with honey.

There are also other methods to check the purity of honey, but they have stirred discussions. The "Ant Test" is a significant exam that is contentious. A group of people claims that ants are drawn to false honey because of their sugar content but are not drawn to real honey because of a natural pest repellent element. Not a lot of data occurs to promote this experiment and, in fact, there is no recognized reason why small animals would prefer processed sugar over actual honey, not forgetting that ants may not always be "accessible" at all locations for honey evaluation. The reason why a sweet liquid is more appealing than another for ants could also be due to other variables such as liquid density, tastes that differ based on the flower type.

If you want to be sure that your honey is of the highest value, you might want to consider purchasing organic honey.

Organic Honey In EU, What Does It Mean?
If you purchased honey labeled organic, that implies that honey is produced from bees that have been supplied only organically grown plants.

Not all that, organic apiculture is performed in a safe setting without intensive farming, the use of synthetic feed and chemicals that are detrimental to bees. If the bees are fed from flowers that have been treated with pesticides or some chemicals, their honey can not be marked organic honey.

Organic honey verifies The region within the 3 km radius of the beehives must be mainly covered by natural plants and/or organic or low-input soil; this is particularly important for plants frequented by bees for feeding (fruit trees, rapeseed, etc.); there must be no significant causes of pollution (agriculture, highways, etc.) in this region. It's a little distinct for the UK, too. They use the norms of the UK Soil Association, in which the radius must be 4 miles and not 3 km as the EU standard.

Hives building also counts; hives must be produced of natural products, without synthetic products such as paints, varnishes, etc.

Feeding of bees is also included, the stock of flowers and seeds must remain adequately strong to endure the lifetime without flowers and honeydew. It is nice to understand that artificial feeding of colonies is permitted, but only with certified organic honey or organically generated sugar products.

Diseases should be avoided as far as feasible. Only legally allowed phytotherapeutic products may be used in preference to allopathic products. For instance, if you want to treat Varroa colony, you are only permitted to use a few authorized goods.

Bees are also important, the genetically modified bee species can not be used. Also, for the first accreditation, all colonies will undergo one year of discussion before their honey could be marked organic honey.

Honey cultivation must be carried out without the use of synthetic repellents. The destruction of bees in comb as a technique connected with the extraction of beekeeping products is forbidden. Be conscious that both the removal of supers and the removal of honey must be recorded.

Long story brief, organic certification verifies that the farm or processing plant complies with organic legislation. Their honey might be expensive, but if it's labeled organic, you can be sure you've bought a real thing.

HOW TO ATTRACT HONEY BEE TO YOUR EMPTY HIVE (BAIT)

A bait hive is an unused hive that has been set up to attract a swarm during the swarming season.

For about a week before the colony swarms, she sends out scout bees to discover a fresh place. Although we should not attempt to humanize bees, there are several requirements that seem appealing to them, which are:-

- Where the bees had resided before.
- Sufficiently large cavity to store enough meat to withstand the winter and possibly the poor season.
- Easily defensible.
- To help ventilation in the shade.
- Dry.
- Where the law lines overlap.

Most beekeepers have seen a bee inspect an unused hive or a bunch of supers during the winter. At first there's only an unusual bee or two, then there's more, often quite upset. This usually occurs for about a week, then either it stops or a swarm emerges. If the exercise had ceased, the swarm could have discovered another house, the beekeeper would have dealt with the colony or gathered the swarm when it had settled.

I've always placed out bait hives, and I've been effective in attracting swarms. I have put them up as follows:-

- Please use a strong ground. I've never had a success with the OMF, probably because the bees don't believe they can protect it.
- An old brood box and crown board that hasn't been washed, so they still have wax and propolis on them. This seems to attract stronger swarms.
- An ancient comb inside if the bait hive is at home, a complete box of ancient comb if it's away from home.
- A tiny door to the bottom or side of the matchbox. Too large and you're going to decrease your likelihood of achievement.
- In the shape of it.

If bees are interested in a bait hive, the first thing you need to do is inspect your own colonies. Even though you may believe they're all right, there's a chance that you've missed superseding cells or queen cells. If they are not your bees, they may originate from another beekeeper or feral colony, so bring the usual precautions to prevent being infected with foul brood.

If the bait hive is at home where I can see what's going on, I'm only using an ancient comb. When the swarm comes, I can shake off the bees, snap the king and complete the brood box with the base. If the bees bring honey / nectar in the comb, it can be burned. If I use a comb away from home where I can't see what's going on, I'm inclined to get a cabinet complete of wild comb, so I fill the box with a comb and bring the danger of having foul brood.

There is a suggestion that bait hives should be placed at or above the head, but that doesn't matter in my view.

I've always had quite a lot of achievement attracting swarms to bait hives, but this has enhanced since I began placing them in locations where the ley lines crossed.

Some beekeepers use swarm lures, but I believe they're cheating and tend to depend on my own abilities.

HOW ARE YOU ENCOURAGING BEES TO CONSTRUCT A COMB?

Bee Honeycomb-Encourage your bees to make it happen!

Honey bees are producing beeswax and combing? It's the normal thing to do you're not supposed to have any problems promoting your bees to produce honeycombs, right? Yeah. Yeah. Perhaps. But there are a number of variables engaged in this vital method of bees. And you can't have too much honeycomb on your side.

Spring is coming and we beekeepers are so thrilled. We're eager to have the bees fill our cabinets with some delightful honey. Bees have a bunch to do, particularly with fresh colonies.

Are you a fresh beekeeper who uses a fresh cabinet with only the beeswax base mounted?

If so, you're likely asking if there's anything you can do to encourage the development of honeycomb bees?

Bees must generate beeswax in order to create honeycomb covers. It's a lot of beeswax.

These individual beeswax cells must be built to maintain eggs, larvae, larvae and meals.

Of course, the honeycomb bee inside the hive has been used for years and years. But the fresh colony is beginning from scratch and ##s a bunch of energy to be used for the manufacturing of wax.

It's not just the fresh bees colonies that we want to start producing wax. There are other factors why beekeepers want to encourage bees to create honeycombs.

It's a nice procedure to turn the ancient comb in the hive. So, sometimes, we want our bees to create some fresh comb images.

And, some individuals like to eat honeycomb, so a new comb is always required.

While we all know that plant nectar (pollen) is the finest nutrition for bees, most beekeepers will eat bees at some point in moment.

Any moment we capture a bee swarm at the end of summer, the survival of the swarm is dubious. They've got a ton of job to do.

Feeding bees 1:1 (equivalent components of plant sugar and water) stimulates the development of wax.

This sugar water resistance is most comparable to plant nectar. It is prevalent practice to supply this percentage of sugar water to colonies that need to be encouraged to construct a comb.

Hungry bees can feed on rainy days, windy days, hot days and night.

Be prepared to feed a large amount, because a jar or two won't do that. Choose a bigger feeder.

How Fast Is The Bees Build Comb?

Our little bees have a scheme. We've got to know how to operate with their natural trends.

The production of beeswax is a essential component of a honey bee lives experience. And you're going to discover beekeepers using distinct words for this method.

Honey bees create draw or create honeycombs of wax generated by the glands on their bodies.

You might hear the word "drawn" or "drawn-out" comb a lot.

Being Southern, it took me a lot to comprehend that individuals were saying "attracted" and not "drone." A Drone is a male bee, and a "drone comb" means something else completely.

Bees would like to construct (or draw) a comb when they need it. But sometimes we beekeepers believe they should be constructing a comb when they seem unwilling to do that.

Sometimes a colony is going to construct a comb very quickly. Other moments, it seems to be taking indefinitely. The bees are unwilling to construct a comb unless they need it.

And, just as important, the colony requires a good population of healthy young bees.

Building Comb Requires Good Beeswax Production In the honey bee colony, the bulk of the population is made up of women workers. These employees are the bees producing beeswax.

Each worker bee has four pairs of wax glands on the bottom of their abdomen. Queen and drones don't have wax glands-they have other employment to do.

Young adult bees are the most successful manufacturers of wax. Most of the honeycomb in your hive is made from bees that are between 10 and 18 days old.

Liquid wax is excreted from wax glands (also known as wax mirrors). It's dried in transparent flakes or scales.

The bee takes the wax scale and utilizes its feet and mouth to mold the wax into hexagon cells. It's not that noteworthy!

What if you've missed a brood cycle due to queen issues, etc., and there's no new bees in the hive?

Adult worker bees of any era may create beeswax.

Older bees that have worked as forgers can become wax manufacturers. They're just not going to be as great as their younger siblings.

Why Don't My Bees Make Honeycombs?

This is a prevalent issue in beekeeping circles. Before I go any further, I would like to tackle the problem of the Queen Exclusive.

The queen excluder is a panel (generally produced of metal) used to maintain the crown out of the honey collection supers.

The aim of this product is to maintain the queen bee out of the honey supers. Her thorax is mildly bigger than the thorax of the worker bees.

They can pass through the screen, and usually the queen can't.

Beekeepers can contend about king excluders for days on end. Are they a nice thing, or are they the worst thing ever?

I just don't want to go there in this post.

But I'm going to say that in all my years of beekeeping, I've only had 1 colony that the workers didn't seem to want to go through the excluder.

All the other colonies have passed through the Exclusive to operate with no problems.

I doubt that having an exclusion device is a deterrent to constructing a comb.

If you want to encourage honey bees to construct a comb, you need to comprehend the dynamics of the hive.

We need three main requirements to be encountered for the construction of the comb. Once these are discussed, we will be able to investigate variables.

Bees Build Comb When You Need It

Remember, the objective of the bee colony and the objective of the beekeeper is not always the same. Most beekeepers want to produce surplus honey.

Honey bee colonies feel the desire to expand in population and potentially to swarm. They want to generate enough honey for the winter.

Honey bees, they're not lazy. But neither are they wasteful. Building honeycomb requires a lot of effort and energy to make the wax and shape the comb.

If they don't need room, you're going to have a difficult time getting them to draw (or construct) a comb.

Usually a colony with a bigger population draws a comb quicker. Not only because they have more youthful bees, but because they think the need to store more honey. They need a bigger pantry to store your baby.

We can encourage the baby bees to construct a comb by supplying them well. This is particularly true when there is a natural nectar flow.

But wait if there's a flow we don't have to care about eating right?

Well, if you've got a feeder on the hive, the bees can eat in the evening, on very humid days and rainy days.

Usually, feeding is going to encourage your baby bees to construct a comb. It's because they're going to need to bring the baby somewhere.

BEES NEED WARM TEMPERATURES IN THE NEST

We understand that bees are eating honey and creating heat in the brood nest. This is the technique used by bees to survive the winter.

And, in the same manner, they can hold the brood nest at the required temperature spectrum.

Bees also need hot temperatures to operate with beeswax. The optimum 95 degrees F is useful for molding beeswax scales to create a comb.

If the outside conditions are very hot or hot, it will be much harder for the bees to construct a comb.

They may not even believe the challenge is worth it. Cold temperatures outside will imply that less food is being carried in.

Honey Bees Need A Lot Of Nectar

Bees must eat big amounts of honey to boost the wax cells. Honey is metabolized to generate wax in fat cells.

The worker bee also used pollen in the first few days of her lives to encourage the growth of fat cells.

Young employees are packed with honey and hung in chains close the building site. This is what we call festooning, and it's really incredible to see.

Festooning bees are also hanging in strings to make honeycombs.

If you have a hive that builds a comb, be very cautious when performing checks. You might be able to see this in the operation of your own hive.

The need for abundance of honey is another reason why feeding may encourage honey bees to construct a comb.

Our excess sugar water will assist to maintain the colony provided with a steady supply of food.

Now we know the three main variables that need to be addressed in order to encourage honey bees to construct a comb.

They must be in need of it, have hot weather and have plenty of meals available.

5 TIPS FOR MAKING HONEYCOMB BEES

1 Optimal Time of Year for Building Comb The time of year can affect how easy it is to get your bees out of the comb. Spring is a moment of natural upheaval.

Colonies feel the desire to boost the population and the potential swarm.

Therefore, it will be simpler to get your bees to create a basis (of building a comb) in the spring.

Try to use your foundation frames early in the year and save a few frames for later emergencies.

Is it difficult to remove the comb later in the year? Of course it's not. If the 3 requirements have been met, the bees will draw the comb.

In my knowledge, however, it is much more hard to encourage honey bees to construct comb after the first of July.

If I need frames constructed out, I'm going to work to get that completed before mid-Summer.

2 Feed Your Bees The Right Sugar Mix To Encourage Wax Production

Feeding your bees will encourage them to remove the comb. Even if the flow of nectar is on, most colonies will use the feeder at night or on rainy days.

If you are feeding your bees for comb construction reasons, use a 1:1 proportion of sugar to water.

This proportion simulates natural nectar and is more probable to result in fresh combs.

No, the honey stored in the hive does not seem to have the same impact. It's a fresh product coming in that stimulates the bees.

But you're not supposed to supply your bees when your honey supers are on, okay? Right! Right! Right! Honey produced from sugar water is not really honey.

However, if you add a super basis, you could feed your bees a little bit. This could offer them a boost to get began.

Understand that you might end up with a little sugar water in your honey super.

3 Baiting Up to Encourage Honey Bees to Build Comb in New Box
If you have a colony that seems unwilling to move up to a fresh box, you may attempt "baiting up."

You bring the complete honeycomb frame out of the snakes ' cabinet and combine it with the base frame.

This may encourage the bees to migrate up to the next box.

There is no harm in doing this. Keep in mind, however, that if you handled mites with that picture in the bees portion of the cabinet, you may not want to eat it.

Beekeepers who use this technique should go home in a couple of weeks and move them back.

4 Swarms Love Building Comb If you capture a honey bee swarm, that's a wonderful chance to get a fresh comb. The employees in the swarm are prepared to make a comb for their fresh house.

They're going to quit the hive with wax glands willing to go! Once the swarms begin to draw the comb, they will often proceed as soon as you supply them.

This may be an chance to draw a few extra frames for subsequent use.

5 Young employees are the best ones to make honeycombs when you are creating divisions or regenerating a nest, maintain in mind that you need a bunch of youthful employees to generate wax at a quick rate.

Put the base frames in the cabinet with the youngest employees to see the finest comb manufacturing.

5 Tips To Encourage Bees To Draw Comb

Time of year Spring is the best way to feed your insects 1:1 sugar water!

Bait up to encourage expansion use swarms to construct comb to maintain a bunch of young adult bees We beekeepers enjoy our bees and want to do everything we can to assist. It's essential to note that we're managing bees, but we're not controlling them.

Our objective is to operate with the natural trends of the colony of honey bees.

If it's Spring, and you've got a good hive of bees, they're supposed to construct a comb when they're prepared.

And, it's not going to harm to make sure you maintain that feeder complete.

If you have excess beeswax after the honey harvest, it can be used for a lot of fun projects.

WARRE HIVE

WHAT IS A WARRE HIVE?

A Warre hive is a vertical top bar hive that uses bars instead of frames, usually with a wooden wedge or guide on the bars from which the bees build their own comb, just like they do in nature.

The Warre (pronounced: WAR-ray) hive is named after its inventor, French monk Abbé Émile Warré. He studied hundreds of different hive styles and settled on this one as the most ideal for bees and beekeeper. His design focused on simplicity, ease of management, and mimicry of honeybees' ideal natural environment. This hive is a vertically stacking top bar hive that incorporates natural comb and the retention of nest scent and heat. We've used these hives since 2008 and find them to be the most hands-off of any hive design.

Why Use A Warre Hive?

Warre hives are ideally suited for the beekeeper looking for a low-cost, low-maintenance hive design. With a Warre hive, there is no need to frequently inspect the colony, purchase an expensive honey extractor or use chemical-laden foundation. Management of Warre hives calls for adding extra boxes to the bottom of the stack (called nadiring), causing comb to be regularly harvested and cycled out of use. This prevents old comb from being reused and therefore ladened with environmental and agricultural chemicals and toxins.

In our mind the Warre hive is the ultimate design for natural, chemical-free beekeeping, and we've had tremendous success with our own Warres with little to no maintenance.

Advantages of using Warre hives

• Ideal for hands-off beekeepers

• Simple management by the box rather than by the comb

• Lighter boxes

• Optional windows

• Foundationless

History Of Warre Hives

Abbe Emile Warre developed the Warre hive over 50 years of research, culminating in what he liked to call "The People's Hive" in the early 1950s. He studied over 300 hive designs, ranging from straw skeps to the modern Langstroth hive, analyzing their ease of use and suitability for honey bees. He focused on simplicity, ease of management, and natural qualities including the building of natural comb (rather than pressed foundation) and the retention of nest scent and heat.

He frowned upon the invasive, tedious micromanagement of individual frames and combs as practiced by most beekeepers in his day. He found it optimal for bees and keeper that to manipulate the hive box by box only a couple times a year, rather than comb by comb every couple weeks. This is key to the beekeeping philosophy that corresponds with Warre hives, and it is a tremendous shift away from the common practices used today. Less bothering of bees means more productivity by and less disturbance to the colony.

Warre would typically add a couple empty boxes to the bottom of the hive in the spring and remove the top boxes from the hive (full of honey) in the fall. This allows for something few other hives offer: The continual cycle of new comb into and old comb out of the hive without the destruction of the precious brood chamber, as each year prior to winter the bees move the excess honey stores to the top of the hive. This removes the pesticide-laden comb from the hive every couple years, making for a healthier, happier colony.

How To Manage A Warre Hive

After using Langstroth, horizontal top bar and Warre hives, we love Warre hives for their simplicity, ease of management and success. We've had more success with Warre hives than any other hive design, and we don't think this is by chance.

BEE INSTALLATION

To install a bee package in a new Warre hive, start with 2 boxes, leaving any extras for holding a feeder. Set the the bottom up box with bars in place, and the top without.Place the queen cage on the bars of the bottom box and remove the cap exposing the hard candy. Shake the worker bees into the top box getting as many into the hive as possible. Then, lean the package near the entrance of the hive to encourage stragglers to move in. Replace the bars in the top box, then place the separator canvas, quilt box, and top it with the roof. Swarms can be installed with the same box configuration, without needing to place the queen cage.

MAINTENANCE HIVE CHECKS

Warre hives were designed to require minimal to no maintenance. Warre simply wanted beekeepers to add empty boxes to the bottom of the hive in the spring and harvest full boxes of honey off the top of the hive in the fall. Thus, Warre hives are meant to be managed by the box rather than by the comb.

Warre hives are meant to be expanded with the nadiring practice, by adding empty boxes to the bottom of the hive. The bees will generally build from the top down, and as their bottom most box becomes 80% full, an empty box can be added underneath. Then, excess honey will be harvested off the top of the hive.

Over the next couple months after installing your bees, monitor their growth by tilting the bottom box forward. If you see the bottom box filling with comb you can add another box or two accordingly. As fall arrives, check the bottom boxes for comb if they are empty you can remove them until the colony is down to 2-3 boxes.

Note: Bees will attach comb to the side walls of hive boxes since they do not have a 4 sided frame. This requires a sharp L-shaped tool to detach the comb. Our ultimate hive tools were designed for this purpose. They may also connect the comb from a top box to the top bars of the box below. This must be disconnected before lifting boxes off one another to avoid comb breakage, and can be done by cracking the propolis seal on all 4 corners, and then running a piano wire or guitar string between the boxes.

HONEY HARVEST

If you have been managing your hive by the box, rather than the individual combs, you will want to harvest by the box as well. This will require clearing bees out of the box to be harvested. You can do this by using a bee escape board. Place the board between the honey supers and the boxes you'd like to leave for the bees. Wait a couple days, especially as nights are getting cooler, and the bees will gradually move out of the supers, unable to find their way back up through the escape!

You can then remove the honey super from the hive and take out the top bars. Cut the comb from the top bars (our Ultimate Hive Tool is great for this), crush it up and strain using a mesh bag. Our Bucket Strainer System streamlines this process for a simple honey harvest.

Note: I don't recommend harvesting any honey in the first season, instead you should leave all of it for the bees and hope for a surplus next season!

Can Bees Recognize You?

Going about their day-to-day business, bees have no need to be able to recognise human faces. Yet in 2005, when Adrian Dyer from Monash University trained the fascinating insects to associate pictures of human faces with tasty sugar snacks, they seemed to be able to do just that. But Martin Giurfa from the Université de Toulouse, France, suspected that that the bees weren't learning to recognise people.

"Because the insects were rewarded with a drop of sugar when they chose human photographs, what they really saw were strange flowers. The important question was what strategy do they use to discriminate between faces," explains Giurfa. Wondering whether the insects might be learning the relative arrangement (configuration) of features on a face, Giurfa contacted Dyer and suggested that they go about systematically testing which features a bee learned to recognise to keep them returning to Dyer's face photos.

The team publish their discovery that bees can learn to recognise the arrangement of human facial features on 29 January 2010 in the Journal of Experimental Biology at http://jeb.biologists.org.

Teaming up with Aurore Avargues-Weber, the team first tested whether the bees could learn to distinguish between simple face-like images. Using faces that were made up of two dots for eyes, a short vertical dash for a nose and a longer horizontal line for a mouth, Avargues-Weber trained individual bees to distinguish between a face where the features were cramped together and another where the features were set apart. Having trained the bee to visit one of the two faces by rewarding it with a weak sugar solution, she tested whether it recognised the pattern by taking away the sugar reward and waiting to see if the bee returned to the correct face. It did.

So the bees could learn to distinguish patterns that were organised like faces, but could they learn to "categorize" faces? Could the insects be trained to classify patterns as face-like versus non-face like, and could they decide that an image that they had not seen before belonged to one class or the other?

To answer these questions, Avargues-Weber trained the bees by showing them five pairs of different images, where one image was always a face and the other a pattern of dots and dashes. Bees were always rewarded with sugar when they visited the face while nothing was offered by the non-face pattern. Having trained the bees that 'face-like' images gave them a reward, she showed the bees a completely fresh pair of images that they had not seen before to see if the bees could pick out the face-like picture. Remarkably they did. The bees were able to learn the face images, not because they know what a face is but because they had learned the relative arrangement and order of the features.

But how robust was the bees' ability to process the "face's" visual information? How would the bees cope with more complex faces? This time the team embedded the stick and dot faces in face-shaped photographs. Would the bees be able to learn the arrangements of the features against the backgrounds yet recognise the same stick and dot face when the face photo was removed? Amazingly the insects did, and when the team tried scrambling real faces by moving the relative positions of the eyes, nose and mouth, the bees no longer recognised the images as faces and treated them like unknown patterns.

So bees do seem to be able to recognise face-like patterns, but this does not mean that they can learn to recognise individual humans. They learn the relative arrangements of features that happen to make up a face like pattern and they may use this strategy to learn about and recognize different objects in their environment.

What is really amazing is that an insect with a microdot-sized brain can handle this type of image analysis when we have entire regions of brain dedicated to the problem. Giurfa explains that if we want to design automatic facial recognition systems, we could learn a lot by using the bees' approach to face recognition.

CONCLUSION

Honeybees provide a wide range of products and facilities (pollination) to human society and the ecosystem. Across the globe, bees are supporting millions of livelihoods while also enriching the ecosystem. Beekeeping is an significant undertaking for rural groups and linked to agricultural and horticultural manufacturing. Although the district of Uttara Kannada has over 60 per cent of its land under forest cover and about 15 per cent under agriculture, the manufacturing of honey is far below its anticipated capacity. The case study found that apiculture could be raised to a much more lucrative undertaking needing small capital investment and high-yield skilled labor compared to other rural employment and poverty reduction programs. We have taken into consideration the static efficiency of apiculture communities, which had played a crucial part in raising consciousness and profitability of science apiculture in their early days, helping individuals to set up apiary facilities in their households and helping to market the products. There are currently several individual entrepreneurs in the district who have noticed the significance of beekeeping, in a consumer society where there is a continuous and growing market demand for honey as a health food, as medicine, for use in confectionaries, in the pharmaceutical industry, and so on.

Books by the same Author:

GREENHOUSE GARDENING FOR BEGINNERS

AQUAPONICS FOR BEGINNERS

BEEKEEPING FOR BEGINNERS

ANDREW MCDEERE

ANDREW MCDEERE · ROSANNE FOX
BACKYARD CHICKENS FOR BEGINNERS

ANDREW MCDEERE
MUSHROOM Cultivation

PSILOCYBIN MAGIC **MUSHROOM**

ANDREW MCDEERE

Search: "Andrew McDeere"

at Amazon

Kind reader,

Thank you very much, I hope you enjoyed the book.

Can I ask you a big favor?

I would be grateful if you would please take a few minutes to leave me a gold star on Amazon.

Thank you again for your support.

Andrew McDeere

Manufactured by Amazon.ca
Bolton, ON